W0043360

 The Bloomsbury Series in Clinical Science

Titles in the series already published:

Bronchoalveolar Mast Cells and Asthma
K. C. Flint

Platelet–Vessel Wall Interactions
Edited by R. Michael Pittilo and Samuel J. Machin

Oxalate Metabolism in Relation to Urinary Stone
Edited by G. A. Rose

Forthcoming titles in the series:

Disorders of Lipoprotein Metabolism
D. J. Betteridge

Immunology of Mycobacterial Infection
G. A. W. Rook

Herpes Simplex Virus
A. Mindel

The Blood Brain Barrier for Clinicians
Alan Crockard and Nicholas Todd

DISEASES IN THE HOMOSEXUAL MALE

Michael W. Adler (ed.)

With 31 Figures

Springer-Verlag
London Berlin Heidelberg New York
Paris Tokyo

Michael W. Adler MD, FRCP, FFCM
Professor of Genito-Urinary Medicine and Honorary Consultant
Physician, Academic Department of Genito-Urinary Medicine,
University College and Middlesex School of Medicine, James Pringle
House, The Middlesex Hospital, London W1N 8AA, UK

Series Editor

Jack Tinker, BSc, FRCS, FRCP, DIC
Director, Intensive Therapy Unit, The Middlesex Hospital, London
W1N 8AA, UK

Cover illustration: (Top left) Dark-ground photomicrograph of
Treponema pallidum. (Top right) CT scan: ring shadow of cerebral
toxoplasmosis with surrounding oedema. (Bottom) Electron
micrograph of an HBV carrier serum showing 42nm Dane particles
and smaller 22nm spherical forms and occasional filaments
($\times$ 44800).

ISBN-13:978-1-4471-1636-3   e-ISBN-13:978-1-4471-1634-9
DOI: 10.1007/978-1-4471-1634-9

British Library Cataloguing in Publication Data
Adler, Michael W. (Michael William)
Diseases in the homosexual male.—(The Bloombury series in clinical science).
1. Male homosexuals. Venereal diseases   I. Adler, Micheal W.
II. Series 616.95′1′008806642
ISBN-13:978-1-4471-1636-3

Library of Congress Cataloging-in-Publication Data
Diseases in the homosexual male/Michael W. Adler (ed.). p. cm.—
(The Bloomsbury series in clinical science)
Includes bibliographies and index.
ISBN-13:978-1-4471-1636-3
1. Sexually transmitted diseases.   2. Gay men—Diseases.
I. Adler, Michael W.   II. Series. [DNLM: 1. Homosexuality.   2. Men.   3. Sexually
Transmitted Diseases—complications.   WC 140 D611] RC200.7.G38D57   1988
616.9′5′08806642—dc19

This work is subject to copyright. All rights are reserved, whether the whole or part of
the material is concerned, specifically the rights of translation, reprinting, re-use of
illustrations, recitation, broadcasting, reproduction on microfilms or in other ways, and
storage in data banks. Duplication of this publication or parts thereof is only permitted
under the provisions of the German Copyright Law of September 9, 1965, in its version
of June 24, 1985, and a copyright fee must always be paid. Violations fall under the
prosecution act of the German Copyright Law.

© Springer-Verlag Berlin Heidelberg 1988
Softcover reprint of the hardcover 1st edition 1988

The use of registered names, trademarks, etc. in this publication does not imply, even in
the absence of a specific statement, that such names are exempt from the relevant
protective laws and regulations and therefore free for general use.

Product Liability. The publisher can give no guarantee for information about drug
dosage and application thereof contained in this book. In every individual case the
respective user must check its accuracy by consulting other pharmaceutical literature.

Typeset by Tradeset Photosetting Limited, Welwyn Garden City, Hertfordshire

2128/3916-543210—Printed on acid-free paper.

Series Editor's Foreword

The Bloomsbury Series in Clinical Science is growing and changing. Its Editorial Board and contributors were all originally selected from, or had links with, the University College and Middlesex School of Medicine. Now, as the Series develops, board members and contributors alike identify with the wider reaches of Bloomsbury and Islington.

The aims of the Series remain, however, to highlight, to review and to record significant areas of research and development in the field of clinical science. All contributors are experts in their particular field and monographs may be the work of a single author or several, guided by individual editors.

Diseases in the Homosexual Male is the third monograph in the Series. Edited by Michael Adler, Professor of Genito-Urinary Medicine at the University College and Middlesex School of Medicine, it presents contributions from a number of distinguished workers with special expertise. AIDS has perhaps highlighted the problem but this monograph illustrates the wider profile and gives witness to the multidisciplinary nature of clinical science.

London, July 1988   Jack Tinker

Preface

It is interesting to reflect that, prior to the advent of the acquired immune deficiency syndrome (AIDS), a book on diseases in homosexual men might have been seen as of little importance. This is not so, of course, as would be attested by those working in the field of sexually transmitted diseases. This book is not about AIDS, even though we have included two clinical chapters on the subject. Instead, it is written to show the wide variety of clinical diseases apart from AIDS that can occur among homosexual men and which can be acquired via the sexual route. The fact that a disease and/or organisms can be spread sexually has often been first realised in homosexual men. This occurred with hepatitis B, protozoal infections, and of course, the human immunodeficiency virus.

I have also felt it important to include two non-clinical chapters on sociological and historical perspectives of homosexuality and AIDS. This has been done to illustrate that "nothing is that new" and that important lessons need to be learnt from history. The lessons show us that the punitive approach to minority groups has never worked.

November 1987   M. W. A.

Contents

1 Male Homosexuality: Cultural Perspectives
J. Weeks 1

2 Bacterial Infections
A. McMillan 15

3 Viral Infections
A. Mindel 41

4 Protozoal Infections
E. Allason-Jones 59

5 Hepatitis
I. V. D. Weller 77

6 Genital Warts
J. D. Oriel 99

7 Syphilis
J. S. Bingham 111

8 AIDS: Epidemiology and Clinical Aspects
M. W. Adler and I. V. D. Weller 129

9 AIDS: Counselling and Support
Part 1: *L. Glover and D. Miller* 163
Part 2: *T. Whitehead* 175

10 AIDS and Homosexuality in Britain: A Historical Perspective
J. Austoker 185

Subject Index 199

Contributors

M. W. Adler
Academic Department of Genito-Urinary Medicine, University
College and Middlesex School of Medicine, James Pringle House,
The Middlesex Hospital, London W1N 8AA

E. Allason-Jones
Academic Department of Genito-Urinary Medicine, University
College and Middlesex School of Medicine, James Pringle House,
The Middlesex Hospital, London W1N 8AA

J. Austoker
Wellcome Unit for the History of Medicine, 45–47 Banbury Road,
Oxford OX2 6PE

J. S. Bingham
Department of Genito-Urinary Medicine, James Pringle House,
The Middlesex Hospital, London W1N 8AA

L. Glover
Department of Genito-Urinary Medicine, James Pringle House,
The Middlesex Hospital, London W1N 8AA

A. McMillan
Department of Genito-Urinary Medicine, Royal Infirmary, Lauriston
Place, Edinburgh EH3 9YW

D. Miller
Academic Department of Genito-Urinary Medicine, University
College and Middlesex School of Medicine, James Pringle House,
The Middlesex Hospital, London W1N 8AA

A. Mindel
Academic Department of Genito-Urinary Medicine, University

College and Middlesex School of Medicine, James Pringle House,
The Middlesex Hospital, London W1N 8AA

J. D. Oriel
Department of Genito-Urinary Medicine, University College
Hospital, London WC1E 6AU

J. Weeks
26 Dresden Road, London N19 3BD

I. V. D. Weller
Academic Department of Genito-Urinary Medicine, University
College and Middlesex School of Medicine, James Pringle House,
The Middlesex Hospital, London W1N 8AA

T. Whitehead
Terrence Higgins Trust, BM AIDS, London WC1N 3XX

Chapter 1

Male Homosexuality: Cultural Perspectives

J. Weeks

This much has become certain: deviancy isn't just a waste product of society, and nor is it intrinsic to the deviant subject. It is, rather, a construction, one which, when analysed, says less and less about the individual deviant and more and more about the society – its structure of power, representation and repression – identifying or demonising him or her.

(Dollimore 1986 p 179)

Introduction

Anyone writing on homosexuality, especially male homosexuality, in the 1980s does so under the shadow of AIDS. This is not because it is a peculiarly "gay plague", as the more scabrous of the popular press would have it; nor, on a world scale, are gay men the chief sufferers of the disease. As a newly recognised, and potentially devastating, phenomenon in the early 1980s AIDS would, moreover, have made a major impact wherever it came from and whoever were identified as its victims. Yet, it is surely undeniable that a major part of the symbolic power of AIDS stems from its association with a still stigmatised sexuality and an unpopular sexual minority in the industrialised countries of the "advanced" West. To that extent, AIDS and homosexuality are today intertwined in a difficult and complicated history.

"History", the black American writer James Baldwin has said, "is the present – *we*, with every breath we take, every move we make, *are* History – and what goes around, comes around (Baldwin 1986 p xiv)". The association between AIDS and homosexuality, and its resultant effect on the way AIDS has been perceived and responded to by everyone from vocal minorities of the fundamentalist Christian Right to prison wardens, theatrical staff, restaurateurs, refuse collectors, undertakers, laboratory technicians, government officials and ministers, is shaped by a living history, by what can be best described as an unfinished revolution in attitudes to homosexuality and lesbian and gay life-styles (Weeks 1985). Although attitudes to homosexuality have liberalised over the past two decades, the subject still trails clouds of fear, prejudice and misapprehension.

History of Sexuality

The quotation at the start of this chapter summarises a general approach that needs to be grasped before the particulars of modern gay life can be understood. Attitudes to homosexuality are socially shaped and historically variable, and cannot be fully explicated without some sense of the complex power relations, social interventions and personal and collective resistances that have shaped the contemporary lesbian and gay male situation. The "history of homosexuality" – if we can so describe a multifarious and highly differentiated series of separate histories – is basically one of a process of definition (by church, state, the law, medicine, psychiatry) and self-definitions (by the people themselves), a struggle over forms of social regulation and public and personal meanings through which modern lesbian and gay male identities have been constructed.

There can be, moreover, no separate account of homosexuality that does not relate it to the social and historical organisation of other sexualities, including those which in the West have assumed the distinction of becoming the norm. The history of sexuality, it has been argued, is a history of a subject in constant flux (Padgug 1979). It is often as much a history of our changing preoccupations about how we should live, how we should enjoy or deny our bodies, as about the past (Weeks 1986 p 21). The careful reconstruction of a homosexual history in recent years is therefore more than an antiquarian curiosity. It is an attempt to understand a complex present. AIDS is one aspect of that present. But its impact has been in part dependent on an even more crucial development, the widespread emergence, since the late 1960s, of forceful lesbian and gay male identities and a strong consciousness of homosexual community. AIDS represents the greatest challenge to that development. It is also a test of how successful gay people have been in creating a firm sense of identity and belonging.

A Living History

Western preoccupation with homosexuality – male homosexuality above all – has a long lineage: even the ancient Greeks saw ethical dilemmas in the conflict between a man's duties to his male lover and his household obligations (Foucault 1986). What has changed significantly over time is the form of that preoccupation, and therefore the nature of its effects. While the ancients were concerned with the problems posed by the conflict between indulgence and restraint, the post-Christian world has been troubled with the relations between "true" and "false" sexualities and who should define them. From the nineteenth century, this has taken the form of an obsessive searching out of "healthy" and "sick", "normal" and "abnormal" sexualities. People who did not, or for whatever reason could not, live up to the prevailing ideal of appropriate erotic behaviour have experienced various forms of regulation and control, and have had to shape their lives, as best they could, within the space allowed them.

There is a statement which polemically sums up much of the recent debate about the roots of modern homosexual consciousness: before the nineteenth century, homosexuality existed, but the homosexual did not. Implicit in this is the argument that, while some form of same-sex activity exists in all cultures, each society ascribes its own meanings to it, and only in a very few cultures has it given rise to a specific social categorisation, and a personal and social sense of self associated with particular sexual practices. In Western societies, the idea that there is a specific type of person called a "homosexual" is of relatively recent origin. This hypothesis has been argued in a number of places (see Weeks 1977; Plummer 1981), and has also been subject to sharp attack. The historical evidence is in fact contradictory, especially over the timing of the development (Foucault 1979; Boswell 1980; Bray 1982; Trumbach 1985). But a number of decisive shifts can be detected that point to the prime significance of changes over the past hundred years or so in shaping the contemporary lesbian and gay male worlds.

In the first place, the late nineteenth century saw a new preoccupation with the detailed regulation of *male* homosexual behaviour. The long-established, broad-brush and spasmodically enforced legislation against sodomy, the sin that was too awful to be named among Christians (and which, formally at least, carried the death penalty in England until 1861, and a life sentence thereafter), made little distinction between anal intercourse of man and man, man and woman, and man and beast. By the second half of the nineteenth century, however, there was a new interest in the detailed specification and legal regulation of other, more minor, breaches of the sexual codes: "gross indecencies", importuning and the like. The result was a formal easing of the capital penalty and an intensified concern with the minor peccadilloes. Oscar Wilde was not imprisoned for "posing as somdomite [sic]", as the Marquess of Queensbury had accused him, let alone for being one, but on relatively minor, but imprisonable, offences with (consenting) working-class lads. The period saw a host of new laws and penalties – concerning, for example, prostitution, obscenity, the age of consent – which marked a critical movement in the emergence of twentieth-century sexual norms.

There was a new emphasis on rules of public decency and private (familial) decorum. In this context male homosexuality in particular became an object of intensified public intervention in the interests of setting the boundary of the normal (Weeks 1981).

Changing Laws

The liberalisation of the laws affecting male homosexuality in Britain after 1967, despite their real importance in creating a new social space, have not fundamentally altered the formal, legal situation regarding homosexuality that we inherited from the nineteenth century. Certain sexual practices in certain situations amongst certain individuals ("consenting adults, over 21, in private") have been decriminalised. Homosexuality itself, however, was not legalised, nor has it been given a status equal with that of heterosexuality.

In a host of examples – from the age of consent (21 for homosexual men, 16 for heterosexual women) to the lack of job security, to difficulties encountered by homosexual parents in child custody cases – formal inequality and discrimination remains. AIDS has accentuated this deeply ingrained prejudice. An opinion poll conducted for London Weekend Television (24 January 1988) suggested that the proportion supporting the legislation of homosexual relations had fallen from 61% in 1985 to 48% in 1988.

Secularisation of Sex

These formal legal changes are closely related to a second major shift, that can best be described as the "secularisation of sex". There has been a progressive detachment of sexual values from religious values, even among many of the religious. This has a long history, but a key element has been the process by which the initiative for judging sexuality passed from the churches to the agents of social and mental hygiene, primarily in the medical profession. Moral and medical matters remain, of course, inextricably linked. You can still be judged simultaneously as both sick and immoral if you stray too far from the norms. Nevertheless, sexuality has become increasingly the province of non-religious experts – in sexology, psychology, welfare services and social policy as well as in medicine itself.

This can be seen in changing ways of conceiving of homosexuality. From being one aspect of a generalised moral state ("sin"), homosexuality increasingly came to be seen as a peculiar psychosocial condition, with its own aetiology, characteristic features and effects.

Natural or Perverse

Pioneering sexologists, from the 1860s onwards, produced ever more complex explanations for, and descriptions of, this strange phenomenon. Was homosexuality a product of corruption or degeneration, congenital or the result of childhood trauma? Was it a natural variation or a perverse deformation? Should it be tolerated or subjected to cure? Havelock Ellis distinguished the "invert" and the "pervert", Freud the "absolute invert", the "amphigenic" and the "contingent". Sometimes later, Clifford Allen distinguished 12 types, including the compulsive, the nervous, the neurotic, the psychotic, the psychopathic and the alcoholic. The last hundred years has seen a sustained attempt by the self-defined "sexual scientists" to search out the causes of homosexuality and to delineate its personal effects.

The "science of sex" has been enormously successful in many ways. By the 1920s even the most stalwart of the traditional moral purity campaigners were prepared to take on board the advice of an Ellis or a Freud. By the 1980s even the most blinkered devotees of the moral Right seek justification for their views by reference to the theories of the latest sexual scientists (the sociobiologists especially). It has even been argued in some quarters that the definitions and categorisations of the sexologists, and the assortment of doctors, psychologists, sociologists, social workers, teachers and the like influenced by them, in a real sense "created" the lesbian and the homosexual man, by so sharply distinguishing the "abnormal" from the "normal", the sheep from the goats, in a sustained effort at social control.

No clear look at the evidence, as we shall see, would fully sustain such an argument. Apart from anything else, the sexologists rarely agreed amongst themselves, nor did they unproblematically attain the status of sanctioned experts. But there was, in their earnest endeavours to seek out the fundamental facts of homosexuality, a common assumption: that by describing a person's sex you were, in an absolutely basic way, telling the *truth* about his or her personality. In the twentieth century, particularly, sex has come to be seen, in a now famous phrase, as the truth of one's being (Foucault 1979).

The belief that you will know the whole person by knowing his or her sexual desires and preferences has had curious effects. On the one hand it lends itself to an approach where the caring professions can find a new responsiveness to particular individual needs. On the other it can become another weapon in the process of labelling and "fixing". If you know someone is "homosexual" then you can draw from the literature, and received prejudice, a host of assumptions by which that person can be placed microscopically, so a gay man can be immediately marked off as "promiscuous", "emotionally unstable", "incapable of full commitment" and so on. This has had dramatic effects in the "professional" approach to homosexuality.

A Disease

Behind this is a less explicit but very widespread assumption that approval or disapproval of particular forms of behaviour is closely related to health or sickness, hygiene or disease. Disease sanctions still largely frame and organise many of our deepest sexual beliefs. It thus becomes easy for certain diseases that are connected, however tangentially in some cases, with sex, such as genital herpes, cervical cancer or more recently AIDS, to appear as punishments, nature's retribution for sexual misdemeanours. In the case of homosexuality, it has over the past couple of hundred years been seen both as a disease itself, and as a cause of disease, a contagion which, unless contained, will not only lead to the innocent young becoming homosexual but will cause even worse diseases.

Given this background, it is not surprising that in the 1980s the response to the emergence of a real disease, like AIDS, could be shaped and coloured by the lingering belief in the "disease of homosexuality". Moral entrepreneurs of the New Right in America and Europe have seized on AIDS as a justification for their moral crusade against homosexuality and the challenges to conventional family life that it is deemed to pose (Weeks 1985). As Mrs Mary Whitehouse put it in an attack on Channel 4 television for thinking of planning a programme on homosexuals: "Over recent years homosexuality has increasingly been represented as being perfectly normal . . . But now the laughing is over . . . Everything possible must be done to protect the public at large from this new and terrible threat [i.e. AIDS]" (quoted in *The Guardian* 11 August 1986).

Identity and Belonging

Modern lesbian and gay identities have emerged in this complex, and hostile, political, moral and ideological environment. Not surprisingly, many men and women have borne the scars of social hostility and internalised self-hatred. Yet there is evidence, over several hundred years, of what has been called a "reverse affirmation" (Foucault 1979), of a sustained effort at self-definition in a constant resistance to the hostile norms. The modern lesbian and gay male identities have been created by this process of self-organisation and self-making, in a continuous history from at least the eighteenth century (Bray 1982; Weeks 1977). The people that the early sexologists described were not figments of their pigeon-holing imaginations, but real people often seeking assistance or advice in an increasingly hostile world.

There was not, of course, a single homosexual identity developing. The homosexual experience then as now was varied, differing along lines of class, geography and even personal need and inclination. Sexuality is far too fluid, far too susceptible to diverse social influences to be understood through a single lens. The intricate patterns of same-sex interaction that develop in such closed societies as prisons illustrates clearly the range of potential roles and identities that emerge (see, for example, Wooden and Parker 1983).

Above all, there is growing evidence for the separate, if inevitably closely related, developments of male homosexual and lesbian identities (Faderman 1985; Freedman et al. 1985). Lesbianism and male homosexuality cannot be straightforwardly linked together as if they stemmed from a single formative cause and revealed a common pathology.

There has been no linear development of a single homosexual sense of self. Yet undoubtedly, behind the apparent and real diversity of needs, aspirations and lifestyles, a strong sense of common identity embracing men and women has developed, predicated on a strong sense of common experience. Modern lesbian and gay identities are not products of any *essential* differences between homosexual and other people, whether rooted in biology, psychology or environmental shaping. Such identities are, however, ways of grappling with a legacy of social hostility, discrimination and prejudice, which have in turn produced a variety of different ways of life.

The very idea of a *sexual* identity is, of course, an ambiguous one. To privilege one's sense of self as a sexual being might involve ignoring one's class, ethnic or professional identities. We all confront a host of possible loyalties, and to which we give priority is dictated by a variety of contingent factors. Yet for many in the contemporary world, especially the sexually marginal, an identity shaped around one's sexuality is absolutely fundamental and necessary, providing a sense of personal unity, social location and even political commitment. Not many may take it upon themselves to say "I am a heterosexual", because it is the taken for granted norm of our culture. But to say "I am gay" or "I am a lesbian" is to make a statement about belonging.

This is not a new phenomenon, as we have seen. Embryonically, at least, individuals have made claims for the priorities of their sexual natures over all other demands for several hundred years, and since the late nineteenth century there have been more or less explicit collective endeavours to affirm the value of homosexual choices (Weeks 1977; Katz 1983). But it is only since the late 1960s that these have become open and forceful, associated with the emergence of new social movements of lesbians and gay men and the growth of gay communities.

Emergence of Movements

Five factors seem to be necessary for these changes to occur: the existence of large numbers in the same situation; geographical concentration; identifiable targets of opposition; sudden events or changes in social position; and an identifiable leadership with readily understood goals (Adam 1978). By the late 1960s all these factors were present in major American centres such as New York and San Francisco, which saw the birth of the gay liberation movement. It was the juncture of the increasingly sophisticated gay and lesbian communities organised around bars and other meeting places, informal networks and a gay press, with a newly politicised movement of "gay liberationists" that provided the energy for the emergence of mass movements in the USA and elsewhere in the 1970s and 1980s.

A social movement can be seen as: "a kind of *ideological hatstand* – a single piece of furniture which, nevertheless, can accommodate a large number and wide variety of hats" (Cohen 1986 p 108). It would be difficult to see in these movements of "sexual minorities" a single ideological affinity, a common social, economic or even ethical location. Yet the lesbian and gay male movements, like feminism, have succeeded in all the major Western countries, and above all in the USA, in transforming the traditional debates on sex by asserting a new claim to self-definition and self-determination on all issues concerning the body and its pleasures.

At the heart of this has been an engagement with, and rejection of, the 'medical model' of homosexuality, and of the right of experts to pin down a person's nature by defining his or her sexuality. The decision of the American Psychiatric Association in 1973 to withdraw homosexuality from its list of diseases was not a result of careful scientific reassessment. It was transparently a result of political lobbying and mobilisation, which reflected a new willingness on the part of homosexual people to break with hostile categorisation (Bayer 1981).

The new social-sexual movements have created, in effect, an alternative public sphere of personal interaction, debates, publications and social and intellectual involvement that have challenged the certainties of what has been described as the "sexual tradition" (Weeks 1985). These movements, which start with "sexuality" and go beyond it in their impact, are by no means unified or coherent in either their means or ends. They have nevertheless introduced a powerful new element into contemporary politics, producing a new "vocabulary of values" (Cohen 1986 p 114) through which individuals construct their understanding of the social world, new constituencies for politicians and businesses to woo, and new "communities of interest", through which a sense of identity is affirmed and reaffirmed.

Such elective communities are not peculiar to sexual minorities. They are in fact an increasingly common feature of the social and political geography of Western industrial countries. Rapid social change has produced new social tensions and antagonisms, forms of domination and resistance, that have shaken traditional political forms and methods. This has given a new political saliency to lesbian and gay male politics in certain areas where numbers of homosexual people are concentrated and where politics are susceptible to interest group influence. Classic examples are provided by San Franscisco and New York. But even in Britain, where the political structure is still to a large degree organised around traditional party groupings and issues, lesbian and gay male activists succeeded in making some gains – for example, in the establishment of lesbian and gay units by some local authorities and by local government supported campaigns against job discrimination, biased sex education and the like. These in turn provoked a backlash, represented for example by Clause 29 of the Local Government Act (1988), banning the intentional promotion of homosexuality by local authorities in the United Kingdom.

The precondition for this new assertion of community rights is a feeling that terms like "gay" and "lesbian" now "connote much more an entire life-style, a way of being in the world which only incidentally involves sexual activity with persons of the same sex" (Bell and Weinberg 1978 p 115). People, as Cohen (1986 p 118) has argued, construct a sense of community "symbolically", making it a resource and repository of meaning, and a major referent of their identity. The lesbian and gay male communities may in some areas have a recognisable geographical location (for example, parts of San Francisco, Greenwich Village, or West Hollywood). Specific physical sites may have a powerful influence as a forcing-house of a sense of identity and community: bars and bath houses have been identified as such (Altman 1982; D'Emilio 1983). But much more important, the "community" exists as an *idea*, embodied in a series of activities (such as gay pride parades, festivals, candlelit vigils for people with AIDS, as well as more intimate and personal involvements) that constantly evoke, recreate and sustain a common belonging, whatever the class, racial, ethnic and gender differences that nevertheless exist and continue to flourish.

Sex and Relationships

Homosexuality has become more than a personal predeliction, or a series of erotic practices: it is for hundreds of thousands of women and men a way of life. Yet this way of life is, inevitably, inextricably concerned with sexual needs and desires and with intimate relationships. It is the "sex" that is stigmatised (as the moral Right can unctuously put it, "we love the sinner but hate the sin"), and the relationships that are invalidated by continuing social hostility.

A word of caution must be put in here. An important tendency among lesbian activists has stressed that lesbianism is more than a sexual definition; indeed it need not have any sexual connotation at all. For them, it is fundamentally about ties of need and sisterhood amongst all women, in which "sexuality" as conventionally defined may be minimal or totally absent (e.g. Rich 1984). For other women, however, as for the majority of gay men, their identities are closely, if not entirely, bound up with their sexual desires (Vance 1984).

The Threat of AIDS

Not surprisingly, the emergence of a major health crisis associated with AIDS can be seen as more than a threat to individual lives and certain sexual practices. It is also, potentially, a threat to only recently achieved positive identities and to a whole way of life (Altman 1986). As the full depth of the health crisis became apparent in the early 1980s, many lesbian and gay male activists resisted what they saw as the threat to their hard fought for sexual spaces. Attempts to close down gay bath houses, as in San Francisco in 1984, when it seemed likely that the spread of AIDS was closely linked with the easy promiscuity that they promoted, met fierce opposition, bitterly dividing the gay community. For some, the right to promiscuous sexual freedom was absolute: sexual promiscuity, it was argued, was the thread that bound the gay male community together (Popert 1982 p 30). For others, the bath houses were an essential social space for the growth of a sense of identity and community (Altman 1986). But for both approaches, the closure of the baths, or their tighter regulation, were seen as new attempts to bring homosexuality once again under authoritarian social control.

For similar reasons, many in the gay community expressed deep suspicion of what was seen as the "remedicalisation" of homosexuality in the wake of the AIDS crisis:

Like helpless mice we have peremptorily, almost inexplicably, relinquished the one power we so long fought for in constructing our modern gay community: the power to determine our own identity. And to whom have we relinquished it? The very authority we wrested it from in a struggle that occupied us for more than a hundred years: the medical profession (Lynch 1982 p 31).

Given these attitudes, it is important to clarify what are the attitudes of the homosexual world to sexuality and relationships in order to disentangle the myths and stereotypes from the lived reality of gay life.

Attitudes to Sexuality and Relationships

Variety of Sexual Needs

The first point that should need no underlining is that gay men, like other men, and women, display a huge variety of sexual needs. Kinsey made the point forcefully some 40 years ago, when he observed with glee the example of two men who might live in the same town, meet at the same place of business, have common social activities, and yet experience enormously different sexual lives: if one had 30 ejaculations a week, the other one ejaculation in 30 years, the difference would be 45 000 times. "This", he wrote, "is the order of the variation which may occur . . ." (Kinsey et al. 1948 p 197).

In a later study, carrying on the Kinsey tradition by looking in detail at the lives of homosexuals in the Bay area of California in the late 1960s, Bell and Weinberg concluded that:

homosexual men and women cannot be sexually stereotyped as either hyperactive or inactive. Rather the amount of sexual activity they reported varied among individuals, with black males reporting more than white males, men reporting more than women, and (with the exception of the black males) younger people of either sex being more sexually active than their older counterparts (Bell and Weinberg 1978 p 72).

The frequency of sexual activity, then, varies enormously, and this is hardly surprising. There is also some evidence that, other things being equal, it is not dissimilar to the male heterosexual pattern. Kinsey and some subsequent commentators noted that, on average, homosexual men had sex less often than their heterosexual brothers – partly because of the lower incidence of settled partnerships – but experienced it with a large number of partners. However, Blumstein and Schwartz in their survey of the sexual lives of different types of American couples concluded that: "gay men have sex more often in the early years of their relationship than any other type of couple. But after 10 years they have sex together far less frequently than married couples . . . Sex with other men balances the declining sex with partners" (Blumstein and Schwartz 1983 p 195). What apparently distinguishes gay men from heterosexual is a greater number of partners or, to use the more loaded term, "promiscuity".

Even here, it is possible to exaggerate. Institutionalised promiscuity (in the form of prostitution) is largely a heterosexual rather than a homosexual phenomenon; and the most notorious form of gay promiscuity, the bath house, has had its heterosexual parallels, as in the Sandstone Sex Commune in California or Plato's Retreat in New York (Talese 1980). On the other hand many gay men do not have multiple partners.

Nevertheless, the 1970s did witness an explosion of what has been described as "public sex" amongst gay men, with the appearance in most of the major American, Australian and European cities (with the partial exception of Britain where the laws remained restrictive) of such facilities as bath houses, backroom bars and public cruising areas where casual, recreational sex with multiple partners became the norm. It was possible to see this development as a depressive challenge to the puritan tradition (see, for example, Hocquenghem 1978), as the forerunner of a Whitman-esque democracy (Altman 1982), or as the harbinger of sexual anarchy, which the American Moral Majority professed to see. Whatever its ultimate meaning (and by the mid 1980s the situation was changing dramatically in the wake of AIDS), it clearly represented some form of de-coupling of sex and intimacy, and a normalisation in a new way of sex as recreation and pleasure. It is too early to say whether this will survive the impact of AIDS. The increasing emphasis in the gay community on avoiding sexual *practices* (such as those, like anal intercourse, which involve the interchange of body fluids) which are likely to encourage the spread of AIDS, and the corresponding emphasis on developing safer sexual activities, suggests that there is less likely to be a wholescale move into celibacy or monogamy than a search for different forms of sexual pleasure.

Sexual Activity and Emotional Partnership

This brings us to the second point: the link between sexual activity and emotional partnerships in the male gay community. There is no reason to think that, other things being equal, gay men find it more difficult to establish intimate emotional and sexual relationships than heterosexuals. Masters and Johnson provided evidence that suggests that in sexual partnerships gay men and lesbian couples tend to be more relaxed, more concerned with a full exchange of pleasure at all levels, less performance oriented and more sensitive to one another's needs than heterosexuals (Masters and Johnson 1979 pp 64–65). They conclude that: "the committed heterosexual couple is handicapped sexually, first by theological covenant, and – a far more important second – by a potentially self-destructive lack of intellectual curiosity about the partner" (Masters and Johnson 1979 p 219).

On the other hand, gay couples tend to ask different types of questions about their relationships from heterosexuals. According to the detailed (Californian) study of gay male couples conducted by McWhirter and Mattison (1984), heterosexual couples did not have to grapple with issues about roles, finances, ownership and social obligations in the same way as gay men do, and were less preoccupied with acceptance by family and neighbours. Heterosexual couples lived with some expectation that their relationships would last "until death do us part", while gay male couples wondered if theirs would survive (McWhirter and Mattison 1984 p 3). On the other hand, gay male relationships were potentially as long lasting, with a patterned development through successful stages of intimacy that is unlikely to be dissimilar to the heterosexual pattern. And where gay male relationships do end, it is necessary to ask if that is any different from the development of a growing heterosexual pattern of divorce and remarriage. As one of Berger's respondents put it: "My sister has gone through three husbands and I have gone through three lovers. The difference is that I have remained good friends with all my ex-lovers" (Berger 1982 p 125). Where gay male relationships do seem to differ substantially, however, is in attitudes to sexual activity outside the partnership. McWhirter and Mattison (1984 p 252) speak of "fidelity without sexual exclusivity" as a normal pattern.

Expectations of fidelity are high among gay men, but they are measured in terms of commitment, not sex. This does not mean that the fact of sex outside the partnership is unproblematic. When the option is exercised, feelings of jealousy, fear of loss and abandonment, and anger can easily erupt. But there seems to be a greater willingness to accept that intimacy and long-term commitment do not necessarily go hand in hand with sexual loyalty.

This is not a peculiarly homosexual preoccupation. The "desacralisation" of sex in the contemporary world, the recognition that sex in itself does not carry a moral weight or obligation, has opened up the way to seeing sex as a matter of choice; what matters is less the nature of the act than the quality of the experience and the context in which it takes place. This is an ambiguous change. On the one hand it opens the way to the full incorporation of sex into the world of commodities and consumerism so that potentially nothing retains any special meaning. Every act can be something to be purchased, consumed and thrown away. This has been one aspect of some feminist criticism of the "sexual revolution". On the other hand, the breakdown of old restraints has opened up the possibility of different forms of relationships, alternative ways of living together, and new possibilities of intimacy. Sexuality and sexual relationships are in a state of unprecedented flux. It is indicative, for example, that Blumstein and Schwartz's (1983) examination of American couples places the traditional pattern as only one amongst four: the married couple, the cohabiting heterosexual couple, the lesbian couple, and the gay couple. In such a situation, gay relationships can come to be seen less as marginal than as one among many forms of living together.

Relationships in the Gay Community

This brings us to a third point that needs underlining: the central place of *relationships* in the gay community. Bruce Voellar (1980) has argued that much of the work of the first decade of the gay rights movement was less concerned with questions of sexuality as such and more with the creation of "surrogate families", composed of other gay persons for mutual support. We may quibble with the "surrogate", because this implies an attempt simply to replicate a traditional family pattern; the evidence does not suggest this. On the other hand, there are clear signs of inventiveness and adaptability in the development of rich and fulfilling friendship networks. As Bell and Weinberg (1978 p 178) observed: "it would appear that homosexual men and women are apt to have more close friends than heterosexuals do".

In fact relationships are often more complex and critical to many people's lives than the conventional term "friendship" would indicate. They range from the sharing of property to mutual support in ill health and misfortune to caring in old age to simple relationships of affection and openness. For many gay people these relationships can be more than substitutes for conventional-couple relationships; they are positive ways of leading a supported and involved life.

Some recent studies of ageing among gay men support this hypothesis. Berger (1982), for example, argued that gays need to learn to master the "crises of independence" earlier than do heterosexuals, as they are forced to risk breaking with family and friends when they come out as gay. This experience, in turn, helps them to negotiate the crises of middle and old age more successfully. Whereas heterosexuals may experience the first crisis of independence in old age, homosexuals have become used to coping without their family of procreation, and are less likely to have become totally dependent on a single relationship: they have in place a more secure network of friends.

This is counterbalanced, of course, by the hazards of prejudice and discrimination. The law does not readily recognise gay partners as next of kin; and in cases of medical emergency, hospitals will often refuse visiting rights to those not connected by blood. Wills, life insurance protection, transfer of property, appropriate recognition of grief and loss, all pose real difficulties for gays facing ill health or death. These problems can become acute in a situation where the illness itself carries a stigma – as AIDS does. Nevertheless, what is remarkable is that, contrary to the traditional pathologising accounts, lesbian and gay people have been able to create, in the face of difficult social difficulties, resilient and enduring relationships that provide the real bonds that knit the gay community together.

A Conclusion

An examination of the recent history of homosexuality casts light on two particular elements. The first is the centrality of collective activity. The emergence of a distinctive sense of lesbian and gay male identity is predicated upon the existence of strong social ties that we have called the lesbian and gay community. This now represents a dense network of loyalties, obligations, friendship ties, social involvements, institutions, meeting places, self-help groupings, newspapers, and so on. It exists symbolically, as a form of identity and belonging. It exists materially as well, in a rich series of activities.

In recent years this collective self-consciousness has been tested to the full. AIDS has had many consequences, but not least among them is its testing of the reserves of strength of the lesbian and gay community. What is surely remarkable is that the community has proved to be resilient and has met the challenge head on. The development of self-help groupings like Gay Men's Health Crisis in New York or the Terrence Higgins Trust or Body Positive in Britain (See Chap.9), all as the result of gay initiative, testify not only to a healthy readiness to confront problems, but also to the strong common ties that have been built up over the previous decades.

The second element that needs stressing follows from this. We suggested earlier that the "modern homosexual" is in a real sense an "invention" of the past couple of hundred years. The word itself did not exist until the later 1860s. The concept barely existed even in the minds of the "experts" till the early part of the century. An awareness of common identity had a slow and uneven spread among homosexually inclined women and men during this century. Yet today we can see a strong homosexual presence that is not simply a product of social change but an active participant in furthering change. This is the significant point.

It would be wrong to suggest that there has been an easy evolution; but even more wrong to believe that history has now reached an end. The response of the lesbian and gay communities to the AIDS crisis suggests not so much that there has been a shuddering setback to the advances of the gay cause. Rather it poses not just a tragic threat but also a challenge and an opportunity. The changes that have been observed in the male gay community in recent years, particularly with the widespread promotion and adoption of "safer sex" practices, suggest that the process of adaptation and development of personal life-styles is continuing. Many gay people have reported, moreover, that the health crisis has strengthened their sense of identity and belonging to their chosen community (Altman 1986). There is no reason to believe that these are isolated responses. The whole history of the past hundred years or so suggests that homosexual people will respond to the opportunity and face the challenge.

References

Adam B (1978) The survival of domination and everyday life. Elsevier, New York

Altman D (1982) The homosexualisation of America, the Americanisation of the homosexual. St Martin's Press, New York

Altman D (1986) AIDS and the new puritanism. Pluto, London

Baldwin J (1986) Evidence of things not seen. Michael Joseph, London

Bayer R (1981) Homosexuality and American psychiatry. Basic Books, New York

Bell AP, Weinberg MS (1978) Homosexualities. A study of diversity among men and women. Mitchell Beazley, London

Berger RM (1982) Gay and gray. The older homosexual man. Alyson Publications, Boston

Blumstein P, Schwartz P (1983) American couples. Money, work, sex. William Morrow, New York

Boswell J (1980) Christianity, social tolerance and homosexuality. University of Chicago Press, Chicago and London

Bray A (1982) Homosexuality in Renaissance England. Gay Men's Press, London

Cohen AP (1986) The symbolic construction of community. Ellis Horwood and Tavistock, Chichester and London

D'Emilio J (1983) Sexual politics, sexual communities. University of Chicago Press, Chicago and London

Dollimore J (1986) The dominant and the deviant: a violent dialectic. Critical Quarterly 28 (1 and 2): 179–192

Faderman L (1985) Surpassing the love of men. Women's Press, London

Foucault M (1979) The history of sexuality, Vol. 1 An introduction. Allen Lane, London

Foucault M (1986) The history of sexuality, Vol. 2 The use of pleasure. Viking, New York and London

Freedman EB, Gelpi BC, Johnson SL, Weston KM (1985) The Lesbian issue. Essays from Signs. University of Chicago Press, Chicago and London

Hocquenghem G (1978) Homosexual desire. Allison and Busby, London

Katz JN (1983) Gay/lesbian almanac. Harper and Row, New York

Kinsey A, Pomeroy W, Martin C (1948) Sexual behavior in the human male. WB Saunders, Philadelphia and London

Lynch M (1982) Living with kaposi's. Body Politic, November: 31

Masters WH, Johnson VE (1979) Homosexuality in perspective. Little, Brown and Co., Boston

McWhirter DP, Mattison AM (1984) The male couple. How relationships develop. Prentice-Hall, Englewood Cliffs

Padgug R (1979) Sexual matters: on conceptualizing sexuality in history. Radical History Review 20: 3–23

Plummer K (ed) (1981) The making of the modern homosexual. Hutchinson, London

Popert K (1982) Public sexuality and social space. Body Politic, July/August: 30

Rich A (1984) Compulsory heterosexuality and lesbian existence. In Snitow A, Stansell C, Thompson S (eds) Desire. The politics of sexuality. Virago, London

Talese G (1980) Thy neighbour's wife. Sex in the world today. Collins, London

Trumbach R (1985) Sodomitical subcultures, sodomitical roles, and the gender revolution of the eighteenth century: the recent historiography. Eighteenth Century Life 9: 109–121

Vance CS (ed.) (1984) Pleasure and danger. Exploring female sexuality. Routledge and Kegan Paul, London and Boston

Voellar B (1980) Society and the gay movement. In Marmor J (ed.) Homosexual behavior. A modern reappraisal. Basic Books, New York, pp 232–252

Weeks J (1977) Coming out. Homosexual politics in Britain from the nineteenth century to the present. Quartet, London

Weeks J (1981) Sex, politics and society. The regulation of sexuality since 1800. Longman, Harlow

Weeks J (1985) Sexuality and its discontents. Meanings, myths and modern sexualities. Routledge and Kegan Paul, London and Boston

Weeks J (1986) Sexuality, Ellis Horwood and Tavistock, Chichester and London

Wooden WS, Parker J (1983) Men behind bars. Sexual exploitation in prison, Da Lapo, New York

Chapter 2

Bacterial Infections

A. McMillan

Introduction

Homosexual men are at risk for the acquisition of bacterial pathogens through oro-genital, oro-anal, peno-insertive and peno-receptive sexual intercourse.

In response to the emergence of the acquired immune deficiency syndrome (AIDS) and the global spread of its causative virus (human immunodeficiency virus, HIV), many homosexual men have reduced the numbers of different sexual partners with whom they have anal intercourse. As a result, the incidence of homosexually acquired syphilis and gonorrhoea in Europe and the United States of America is falling (Judson 1983; Weller et al. 1984; Poulsen and Ullman 1985). It should be noted, however, that the sexual activities that predispose to bacterial infection are still practised and physicians who undertake the management of homosexual men should be familiar with their clinical features, diagnosis and treatment. These infections will now be considered separately, but it should be stressed that concurrent infections are common (Quinn et al. 1983). Syphilis is considered in Chap. 7.

Neisseria gonorrhoeae

Aetiology

Asymptomatic urethral gonorrhoea is uncommon amongst homosexual men (Handsfield et al. 1980). In the uncomplicated case, *Neisseria gonorrhoeae* infects the urethra, rectum and pharynx (Table 2.1). This finding has been explained by the relative rarity of infection of this group of men with strains of *N. gonorrhoeae* that require arginine (Arg), hypoxanthine (Hyx) and uracil (Ura) for growth, the auxotype that is most commonly isolated from men with symptomless urethral infection (Crawford et al. 1977). Givan and Jaeger (1986) confirmed the rarity of infection of

Table 2.1. Distribution of sites infected with *Neisseria gonorrhoeae* in homosexual men

Percentage of men with gonorrhoea yielding positive results on culture of material from:							No. of men infected with *Neisseria gonorrhoeae* at any site	Reference
Urethra only	Rectum only	Pharynx only	Urethra and pharynx	Urethra and rectum	Pharynx and rectum	Urethra, pharynx and rectum		
53	33	5	1	7	1	1	278	McMillan and Young (1978)
35	28	2	NS	NS	9	NS	54	Austin et al. (1978)
42	32	6	3	9	7	2	270	Janda et al. (1980)
39	27	4	7	15	6	2	85	Handsfield et al. (1980)
35	38	4	3	11	6	3	405	Carlson and Haley (1984)

NS, not stated.

the rectum, anorectum and pharynx of homosexuals with Arg^- (ornithine $(Orn)^-$), Hyx^- Ura^-, auxotypes. They also showed that the auxotype distribution of urethral isolates from these men was different from that of *N. gonorrhoeae* cultured from heterosexual men. Of urethral isolates from homosexuals, 92% were either proto-trophic (non-requiring), required proline (Pro^-) or ornithine (Orn^-). The preva-lence of prototrophic and Pro^- auxotypes amongst gonococcal strains cultured from the urethras of heterosexual men was lower and the prevalence of Pro^-, Pro^-, Cit^- (citrulline), $RUra^-$, Orn^-, Ura^-, and Hyx^-, was higher than in isolates from homosexuals. Similar results were obtained when the auxotypes of isolates from the pharynx and anorectum of women and homosexual men were compared.

Using a coagglutination system for the detection of specific antigens of Protein I (major outer-membrane problem), gonococcal strains can be classified into two main serogroups: WI and WII/WIII. As these serogroups are stable during natural trans-mission, such a classification system has been used in epidemiological studies. From Table 2.2 it may be seen that there is a lower prevalence of WI strains and a higher prevalence of WII strains isolated from various sites in homosexual men. Amongst these men, serogroup WII has predominated, regardless of the site from which the strain has been isolated (Reid and Young 1984). Interestingly, there is a strong corre-lation between WI serogroup and the auxotypes Arg^-, Hyx^-, and Ura^- (see above). β-Lactamase-producing strains of *N. gonorrhoeae* (PPNG) are isolated uncommonly from homosexual men. Only 3% of 1380 PPNG strains in the Netherlands during 1982 had been acquired homosexually (Ansink-Schipper et al. 1984). In the USA between 1976 and 1980, only 1.2% of 506 β-lactamase-producing strains had been homosexually acquired (Jaffe et al. 1981).

Amongst non-β-lactamase-producing *N. gonorrhoeae*, WI and WIII strains are the most and least susceptible to benzylpenicillin, respectively. Differences in the penicil-lin sensitivity of WII strains acquired in different geographical areas are apparent. Although 45% of strains acquired in Stockholm had a minimum inhibitory concen-tration (MIC) of benzylpenicillin of < 0.06 mg/l, none of 17 non-PPNG strains from Thailand was inhibited by this concentration of drug (Bygdeman et al. 1981). Similar

Table 2.2. Distribution of serogroups WI and WII/WIII among isolates of *Neisseria gonorrhoeae* from homosexual and heterosexual men

Percentage of strains (total number of isolates) with serogroups				Reference
WI from		WII/WIII from		
Homosexual men	Heterosexual men	Homosexual men	Heterosexual men	
15 (39)	37 (76)[a]	85 (39)	58 (76)[a]	Bygdeman (1981)
23 (57)	44 (93)	77 (57)	55 (93)	Morse et al. (1982)
10 (59)	66 (56)	90 (59)	34 (56)	Reid and Young (1984)

[a]Data from previously reported study from same city (Stockholm); 13 of the 76 infected men were homosexual (Bygdeman et al. 1981).

differences in susceptibility of the serogroups to doxycycline and cefuroxime have been noted (Bygdeman 1987). WI strains are more sensitive to doxycycline and cefuroxime than is WII, and locally acquired strains of the latter serogroup are more susceptible than those from Thailand.

These data, and the interesting observation that WI isolates from homosexual men are more resistant to antibiotics than are WI strains from heterosexual men, suggest heterogeneity of the serogroups. Using polyclonal antibodies, there was difficulty in resolving subgroups within the WII/WIII serogroups. Monoclonal antibodies against epitopes on Protein I have overcome some of these difficulties. Using panels of monoclonal antibodies in coagglutination reactions, a pattern of reactivity of each strain is recognised – the serovar. After a capital A or B for antibodies that identify the two principal Protein I molecules that are possessed by WI and WII/WIII strains, respectively, lower-case letters denote the reagents with which that strain reacts.

Backman et al. (1985) noted that the WI and WII/WIII serovars of gonococcal isolates from women and heterosexual men in Stockholm were more often Aedgih and Baik, respectively, than those from homosexual men. Ae and Bacek (corresponding to Bak and Bacejk), however, were more frequently identified in the latter group. Similarly, Bacejk serovar was more frequently found amongst isolates from homosexual men in Australia and New Zealand (Bygdeman 1987). Concurrent infection with two or more different serovars was more common amongst homosexual men than amongst women and heterosexual men in Stockholm (Backman et al. 1985) and strains of different serovars were identified more frequently amongst homosexual than amongst heterosexual couples (Bygdeman 1987).

Antimicrobial resistance has correlated with certain WII/III serovars. In a study of the susceptibility of 85 isolates to thiamphenicol and rifampicin, strains with decreased susceptibility to those drugs reacted with the monoclonal antibody Bg (Bygdeman et al. 1984). Similarly, the MIC 50% of benzylpenicillin for strains reacting with Bg was significantly higher than for strains that did not react with this antibody (Bygdeman 1987); all Bg-reacting strains showed reduced sensitivity to benzylpenicillin (MIC > 0.125 mg/l). Decreased susceptibility to rosoxacin of serovar Bacek (see above) was noted by Ruden et al. (1985); the MIC 50% for the serovar Ae (see above) is significantly higher than for other WI serovars (Bygdeman 1987).

Strains of *N. gonorrhoeae* that infect homosexual men, then, are likely to be of serogroup WII and to be serovars that show reduced sensitivity to antimicrobial agents. This may explain the well-recognised difficulty in treating anorectal infection in this group of men (see below). The reasons for the increased prevalence of such strains, however, are uncertain. There is a genetic linkage between serogroup specificity and antimicrobial susceptibility (Bygdeman et al. 1982).

The genetic locus *mtr* confers resistance to hydrophobic dyes, detergents and antibiotics. Morse et al. (1982) showed that *mtr* strains were more prevalent amongst homosexual than amongst heterosexual men and that 17 of the 19 strains studied were of serogroup WII. Fagan (1985) showed that the prevalence of *mtr* strains isolated from the rectums of homosexual men was much higher than that from the urethras of heterosexual men. As these strains were more resistant than non-*mtr* strains to growth inhibition by faecal lipids, Morse et al. (1982) postulated that selective pressures in the rectum, a site rich in hydrophobic molecules, favoured the emergence of resistant strains.

The mechanism of this resistance to hydrophobic molecules is uncertain and a discussion of this issue lies outwith the scope of this chapter.

Clinical Features

Urethral Infection. As indicated above, symptomless urethral gonorrhoea in homosexual men is uncommon. Most patients complain of a urethral discharge and dysuria of variable severity, symptoms that usually develop within two to four days of sexual contact with an infected partner. A mucopurulent discharge is noted at the urethral meatus. Epididymitis is the most common complication of untreated urethral infection. Other rare complications include abscess formation of the parafrenal (Tyson's) glands, periurethral cellulitis and abscesses, inflammation and abscess formation in Cowper's glands, prostatitis and seminal vesiculitis.

Rectal Infection. Although the prevalence of symptoms in men with rectal gonorrhoea has varied widely from series to series (McMillan and Young 1978; Fluker et al. 1980; Lebedeff and Hochman 1980), it is clear that asymptomatic infections are common (Table 2.3). Symptoms include a mucopurulent anal discharge that often streaks the surface of the stools, mild anal bleeding, perianal discomfort, pruritus ani and, rarely, severe pain and tenesmus. These features, however, are not specific and many homosexual men with no identifiable infection give a history of symptoms referrable to the anal region (Owen and Hill 1972; McMillan and Young 1978). During proctoscopy, the rectal mucosa appears normal in over 80% of infected men (McMillan et al. 1983). When there are signs of proctitis – loss of normal vascular pattern, oedema, contact bleeding, presence of mucopus in the lumen of the rectum – these are generally confined to the distal 10 cm of the rectum (McMillan et al. 1983).

Although mild inflammatory cell infiltration of the lamina propria is noted in about 20% of rectal biopsies from non-infected men, similar changes are found in more than 40% of patients with rectal gonorrhoea (McMillan et al. 1983). The histological appearance, however, is not specific, the most common finding being an increased number of lymphocytes and plasma cells within the lamina propria; occasionally a more acute proctitis with a polymorphonuclear leucocyte infiltration is seen.

Perianal abscess formation is apparently an uncommon complication of untreated rectal gonorrhoea. In a personal series of 500 infected men, only one developed this complication. As noted above, disseminated gonococcal infection is rare amongst homosexual men, but a case with systemic spread from the rectum was described by Holmes et al. (1971).

Pharyngeal Infection. Although most patients with gonococcal colonisation of the pharynx are symptomless (Osborne and Grubin 1979; Janda et al. 1980), some men complain of soreness or discomfort in the throat, sometimes with difficulty in swallowing and mild pyrexia (Fiumara 1979). These symptoms, however, are nonspecific. Osborne and Grubin (1979) reported no correlation between the results of culture for *N. gonorrhoeae* and symptoms. Wiesner et al. (1973) noted a correlation between pharyngeal symptoms and the practice of fellatio, but not with gonococcal infection.

Table 2.3. Prevalence of anorectal symptoms in men with rectal gonorrhoea

Percentage of men with anorectal symptoms (Number of men with rectal gonorrhoea studied)	Reference
58 (19)	Owen and Hill (1972)
30 (60)	McMillan and Young (1978)
66 (73)	Fluker et al. (1980)
82 (554)	Lebedeff and Hochman (1980)
38 (134)	Janda et al. (1980)
62 (26)	Munday et al. (1981a)

Diagnosis

The definitive diagnosis of gonorrhoea at any site in the body rests with culture of the organism from that site. A presumptive diagnosis of urethral gonorrhoea can be made by microscopical examination of Gram-stained smears of exudate. The predictive value of a positive smear (percentage of cases with positive smears who have positive cultures) is about 91% and the sensitivity (percentage of cases with positive cultures who have positive smears) is about 94% (Goh et al. 1985). Although the predictive value of a positive smear of rectal exudate is about 87%, the sensitivity is only about 58% (McMillan and Young 1978). It is, however, good practice to examine Gram-stained smears of rectal material so that, if typical organisms are seen, treatment of some patients can be initiated before culture results are available. As the sensitivity of smear microscopy of rectal exudate is low and as other neisseriae can colonise the rectum (see below), culture and the characterisation of any isolate by biochemical or immunological tests is essential for diagnosis. Material for culture can be obtained on cotton-wool-tipped applicator sticks passed through a proctoscope or, blindly, through the anal canal; both methods are equally reliable (Deheragoda 1977). As about 7% of men with rectal gonorrhoea may not be identified by culture on only one occasion (McMillan and Young 1978), the author considers that material for culture should be taken again several days after the first test, if this yielded negative results.

In the diagnosis of pharyngeal gonorrhoea, Gram-smear microscopy is useless. Culture on a selective medium is essential and, as at least 25% of infections will not be detected by the taking of material for culture on only one occasion, repeat testing is necessary (McMillan and Young 1978).

Treatment

Urethral Infection. The treatment of urethral gonorrhoea in homosexual men is similar to that in heterosexual patients (Table 2.4). The choice of antimicrobial agent

Table 2.4. Results of treatment of men with urethral infection with non-β-lactamase-producing *Neisseria gonorrhoeae*

Antimicrobial agent	Dosage	Percentage of men cured (Number treated)	Reference
Penicillins			
Benzylpenicillin	3 g i.m.[a]	98 (191)	Thin (1974)
Procaine penicillin	1.68 g i.m.[a]	97 (109)	Taylor and Seth (1975)
Ampicillin	2 g orally[a]	94 (112)	Taylor and Seth (1975)
Ampicillin	3 g orally[a]	95 (115)	Taylor and Seth (1975)
Talampicillin	1.48 g orally[a]	98 (245)	Price et al. (1977)
Amoxicillin	3 g orally	99 (81)	Price and Fluker (1975)
Amoxicillin and clavulanic acid	3 g/125 mg orally[a]	98 (58)	Lawrence and Shanson (1985)
Cephalosporins			
Ceftizoxime	0.5 g i.m.	100 (69)	Spencer et al. (1984)
Ceftriaxone	125 mg i.m.	100 (26)	
Cefuroxime	750 mg i.m.[a]	95 (100)	Morrison et al. (1980)
Cefuroxime axetil	1.5 g orally[a]	100 (50)	Wanas and Williams (1986)
Spectinomycin	2 g i.m.	99 (110)	Porter and Rutherford (1977)
Aztreonam	1 g i.m.	100 (64)	Evans et al. (1986)

i.m., intramuscular injection.
[a]Given with probenecid in an oral dosage of 1 g.

depends on the susceptibility of prevailing strains to that drug but helpful guidelines have been published by the Centers for Disease Control, USA (1982). Although strains of *N. gonorrhoeae* isolated from homosexual men show reduced sensitivity to various chemotherapeutic agents, these treatment regimens are generally satisfactory.

Rectal Infection. It is common clinical experience that rectal gonorrhoea in men is more difficult to treat than urethral infection. Results of single-dose treatment with penicillin have varied (Table 2.5) but unacceptably low cure rates have been reported. Failure rates with the tetracyclines and cotrimoxazole and reports of adverse reactions in HIV-infected individuals to the latter drug (Gordin et al. 1984) limit their use. In most published studies, spectinomycin has proved useful and is well tolerated by most patients. Experience in the treatment of rectal gonorrhoea with the new cephalosporins is limited, but encouraging results are being reported (Table 2.5). As about 5% of patients with penicillin hypersensitivity show a similar reaction to cephalosporins, those agents are best avoided in individuals with such a history. As only a few men with rectal gonorrhoea have been treated with newer antimicrobial agents such as rosoxacin, conclusions about their efficacy cannot be drawn.

Treatment schedules for rectal infection with β-lactamase producing *N. gonorrhoeae* have not been reported, but it is likely that spectinomycin and the second and third generation cephalosporins will be effective.

Complicated infections require admission to hospital and treatment with intravenous and intramuscular antimicrobial agents. Perianal abscesses may require surgical drainage.

Pharyngeal Infection. Although spontaneous resolution of pharyngeal gonorrhoea has been reported (Wallin and Siegel 1979), most physicians treat infection at this

Table 2.5. Results of treatment of men with rectal infection with non-β-lactamase-producing strains of *Neisseria gonorrhoeae*

Antimicrobial agent	Dosage	Percentage of men cured (Number treated)	Reference
Penicillins			
Procaine penicillin	0.36 g i.m. stat.	64 (36)	Scott and Stone (1966)
Procaine penicillin	1.08 g i.m. stat.	73 (96)	Fluker and Boulton Hewitt (1970)
Procaine penicillin	2.88 g i.m. stat.	65 (20)	Owen and Hill (1972)
Procaine penicillin	2.88 g i.m. stat.	96.7 (451)	Sands (1980)
Procaine penicillin	2.88 g i.m. stat.	96 (99)	Lebedeff and Hochman (1980)
Ampicillin	3.5 g orally with probenecid stat.	85 (27)	Sands (1980)
Ampicillin	3.5 g orally with probenecid stat.	95 (110)	Lebedeff and Hochman (1980)
Ampicillin	4.5 g orally with probenecid stat.	95 (100)	Lebedeff and Hochman (1980)
Ampicillin	250 mg q.i.d. for 5 days stat.	94 (103)	John and Jefferiss (1973)
Amoxicillin with clavulanic acid	3 g and 250 mg orally stat.	100 (17)	Silva et al. (1984)
Amoxicillin with potassium clavulanate	3 g and 250 mg orally stat.	95 (19)	Munday et al. (1985b)
Cephalosporins			
Cefuroxime	1.5 g i.m. stat.	83 (24)	Stolz et al. (1984)
Cefotaxime	1 g i.m. stat.	100 (20)	Stolz et al. (1984)
Cefotaxime	1 g i.m. stat.	91 (11)	Handsfield and Holmes (1981)
Ceftriaxone	125 mg i.m. stat.	100 (52)	Judson et al. (1985)
Cefoxitin	2 g i.m. stat.	95 (20)	Greaves et al. (1983)
Tetracyclines			
Tetracycline	3 g in divided doses	80 (10)	Owen and Hill (1972)
Tetracycline	9.5 g orally in divided doses	85 (98)	Lebedeff and Hochman (1980)
Tetracycline	2 g orally daily for 4.5 days	86 (97)	Sands (1980)
Cotrimoxazole (trimethoprim 80 mg, sulphamethoxazole 400 mg per tablet)	2 tablets orally twice daily for 7 days	88 (66)	Waugh (1970)
Kanamycin	2 g i.m. stat.	84 (90)	Fluker and Boulton Hewitt (1970)
Spectinomycin	4 g i.m. stat.	100 (127)	Fiumara (1978)
Spectinomycin	2 g i.m. stat.	96 (56)	Fluker et al. (1980)
Spectinomycin	2 g i.m. stat.	100 (25)	Sands (1980)
Spectinomycin	2 g i.m. stat.	95 (22)	Handsfield and Murphy (1983)
Moxalactam	1 g i.m. stat.	95 (21)	Handsfield (1983)

i.m., intramuscular injection; stat., statim; q.i.d., four times per day.

Table 2.6. Results of treatment of pharyngeal gonorrhoea caused by non-β-lactamase-producing *Neisseria gonorrhoeae*

Antimicrobial agent	Dosage	Percentage of treatment failures (Number of patients treated)	Reference
Penicillins			
Ampicillin	2 g stat. orally with probenecid 1 g	47 (94)	Odegaard and Gundersen (1973)
Ampicillin	1 g i.m. followed 4 h later by 2 g orally	71 (14)	Stolz and Schuller (1974)
Ampicillin	2 g stat. orally with probenecid 1 g	54 (78)	Hallqvist and Lindgren (1975)
Ampicillin	1 g orally q.i.d. for 3 days	3 (38)	Hallqvist and Lindgren (1975)
Ampicillin	3.5 g and probenecid 1 g stat. then 0.5 g q.i.d for 2 days	4 (77)	Dicaprio et al. (1978)
Procaine penicillin	2.88 g i.m. stat.	3 (36)	Wiesner et al. (1973)
Benzyl penicillin	3 g i.m. with probenecid 1 g stat.	50 (20)	Bro-Jorgensen and Jensen (1973)
Procaine penicillin	2.88 g i.m. stat.	14 (93)	Fiumara (1979)
Tetracyclines			
Tetracycline	2 g orally q.i.d. for 1.25 days	0 (19)	Wiesner et al. (1973)
Cephalosporins			
Cefuroxime	1.5 g i.m. stat.	29 (14)	Stolz et al. (1984)
Cefuroxime	1.5 g i.m. stat. with 1 g probenecid	46 (13)	Graudal et al. (1985)
Cefotaxime	1 g i.m. stat.	9 (11)	Stolz et al. (1984)
Ceftriaxone	125 mg i.m. stat.	6 (32)	Judson et al. (1985)
Spectinomycin	4.0 g i.m. stat.	54 (13)	Wiesner et al. (1973)
	4.0 g i.m. stat.	29 (24)	Karney et al. (1977)
	2.0 g i.m. stat.	43 (14)	Judson et al. (1985)
Cotrimoxazole each tablet containing tri- methoprim 80 mg in sulphameth- oxazole 400 mg	2 tablets orally twice daily for 7 days	3 (29)	Bro-Jorgensen and Jensen (1973)

i.m., intramuscular injection; stat. statim; q.i.d., four times per day.

site because of the risk of transmission of infection through fellatio and, rarely, of disseminated disease within the individual (Metzger 1970).

In general, reports on treatment of pharyngeal gonorrhoea do not differentiate between homosexually and heterosexually acquired infection but the high failure rates with the drugs and doses used in the treatment of uncomplicated genital infections are evident (Table 2.6). Treatment regimens for pharyngeal infection with β-lactamase-producing gonococci have not been established, but daily injections of 0.75 to 4.5 g cefuroxime for three to seven days have proved effective (Lindberg et al. 1982).

Contact tracing of infected individuals is essential and should be undertaken by skilled staff. Microbiological investigations should be repeated after treatment of rectal and pharyngeal gonorrhoea. As in the diagnosis of infection at these sites, the author prefers to have negative culture results on each of two consecutive occasions before declaring cure.

Table 2.7. Frequency of isolation of *Neisseria meningitidis* from the pharynx and anogenital tract

Site sampled	Percentage (Number of specimens cultured) of patients from whom *N. meningitidis* is isolated		Reference
	Homosexual men	Heterosexual men	
Oropharynx	42.5 (815)	NS	Janda et al. (1980)
	35.6 (362)	NS	Salit and Frasch (1982)
Urethra	0.7 (815)	NS	Janda et al. (1980)
	0.4 (669)	NS	Judson et al. (1978)
	1.1 (1363)	0.2 (1720)	Carlson et al. (1980)
	0.3 (357)	NS	Salit and Frasch (1982)
Anorectum	2.0 (815)	NS	Janda et al. (1980)
	2.1 (731)	NS	Judson et al. (1978)
	2.4 (1401)	NS	Carlson et al. (1980)
	2.5 (316)	NS	Salit and Frasch (1982)

NS., not stated.

Neisseria meningitidis

Pharyngeal Colonisation (Table 2.7). *Neisseria meningitidis* is the most common species of *Neisseria* that colonises the oropharynx. There is a seasonal and geographical variation in the prevalence of infection, the carriage rate being highest in the winter months. The organism is isolated from the oropharynx of homosexual men more frequently than from that of heterosexuals (Faur et al. 1981; Janda et al. 1980). Many isolates are ungroupable but, of those that have been serogrouped, serogroup B is the most common (Faur et al. 1981; Janda et al. 1983). The serogroup of the strain of *N. meningitidis* that colonises the oropharynx of homosexual men will, however, depend on the prevalence of that serogroup in the general population. At present there is no evidence that particular serogroups are more likely to be acquired homosexually. Protein and lipoprotein serotypes occur within serogroups B, C, Y and W135, and it would be interesting to know if certain of these serotypes are found with increased frequency in the pharynx or at extrapharyngeal sites in homosexual men.

Simultaneous infection of the pharynx with *N. meningitidis* and *N. gonorrhoeae* is uncommon (Faur et al. 1981; Salit and Frasch 1982) and it has been suggested that meningococci may produce factors that inhibit the growth of the gonococcus. Indeed such activity has been shown in vitro (Bisaillon et al. 1984). Natural resistance to gonococcal pharyngitis, however, cannot be explained solely on this basis and other factors such as competition for essential nutrients may be important.

Urethral Colonisation (Table 2.7)

Neisseria meningitidis is cultured occasionally from the urethras of homosexual men and even more uncommonly from heterosexuals (Carlson et al. 1980). Although the urethra may rarely be the only site of meningococcal infection, the pharynx is usually infected also. The serogroup and antibiotic resistance patterns of the isolates may, however, be different (Janda et al. 1983).

The urethral carriage of *N. meningitidis* may be symptomless or associated with the clinical features of urethritis (Givan et al. 1977; Miller et al. 1979; Faur et al. 1981;

Janda et al. 1983). In assigning a causation for *N. meningitidis* in reports, it should be emphasized that in many reported cases concurrent infection with urethral pathogens such as *Chlamydia trachomatis* has not been excluded.

Anorectal Colonisation

Although the isolation rate of *N. meningitidis* from the rectums of homosexual men is relatively low (Table 2.7), within recent years the proportion of neisserial isolates that are identified as meningococci is increasing. In 1985 in Edinburgh, 16% of rectal neisserial isolates were *N. meningitidis*; the isolation rate in 1980 was only 8%. This finding probably reflects the decreasing incidence of rectal gonorrhoea amongst homosexual men.

In some patients there is concurrent oropharyngeal and rectal infection, sometimes with the same antibiogram and serogroups (Janda et al. 1983). Often, however, the pharyngeal and rectal isolates are different and rectal infection alone is not uncommon. This suggests that infection at this site can result from orogenital followed by peno-receptive anal intercourse or directly from oro-anal contact.

Although some men with rectal meningococcal infection complain of anorectal discharge, perianal discomfort or burning, the majority of patients are symptomless (Judson et al. 1978; Faur et al. 1981) and have no histological evidence of proctitis. In most reported studies, concurrent infection with other pathogens has not been excluded and much remains to be learned about the pathogenicity of *N. meningitidis* for the intestinal tract.

Chlamydial Infections

The genus *Chlamydia* contains two species: *C. trachomatis* and *C. psittaci*. The latter species is associated with pneumonia and, with the exception of a newly identified strain (TWAR) that has been isolated from young patients with acute respiratory tract infections and may be spread from human to human, is acquired from birds. *Chlamydia trachomatis* contains two biovars: *trachomatis* and *lymphogranuloma venereum*. By microimmunofluorescence, biovar *trachomatis* is subdivided into 12 serovars: A, B, Ba and C are associated with hyperendemic trachoma and serovars D to K with oculogenital infection. Serovars L1, L2 and L3 are associated with lymphogranuloma venereum (LGV).

Clinical Aspects

Urethral Infection

Chlamydiae are now well recognised as important causes of non-gonococcal and post-gonococcal urethritis in heterosexual men (Taylor-Robinson and Thomas 1980), being isolated from about 50% of patients. Their importance in urethral disease in homosexual men is less certain. Bowie et al. (1978) failed to culture these organisms from the urethras of 18 homosexual men with untreated gonorrhoea and Stamm et al. (1984) showed that the prevalence of urethral chlamydial infection in homosexuals was lower than in heterosexuals (5% versus 15%). Chlamydiae however, were isolated from this site with equal frequency from the homosexual and heterosexual patients studied by two British groups (Oriel et al. 1976; Munday et al. 1981b). Unlike Stamm (1984), who noted a higher prevalence of serum chlamydial antibodies amongst homosexual men than amongst heterosexuals (52% and 46%, respectively), a low prevalence (26%) of IgG antibodies was noted by Munday et al. (1981a) working in London and by A. McMillan in Edinburgh (unpublished work). In general, these prevalence rates are lower than those found in heterosexual men attending sexually transmitted disease (STD) clinics.

Rectal Infection

Chlamydia trachomatis was isolated for the first time from the rectums of two homosexual men by Goldmeier and Darougar (1977). In patients selected from STD clinics the prevalence of rectal chlamydial infection in men, however, is low (Table 2.8). Rompalo et al. (1986) noted that the highest age-specific prevalence occurred in the 15 to 20 year old age group.

Table 2.8. Isolation of *Chlamydia trachomatis* from the rectums of homosexual men who attended STD clinics

Number of patients (%) from whose rectums *C. trachomatis* was isolated/ number of patients tested	Reference
6 (4)/150	McMillan et al. (1981)
10 (6)/180	Munday et al. (1981a)
24 (8)/288	Stamm et al. (1982)
21 (7)/309	Munday et al. (1985a)
87 (6)/1429	Rompalo et al. (1986)

Rectal infection with the oculogenital serovars of *C. trachomatis* is often symptomless; fewer than 40% of infected men have clinical features of proctitis (Munday et al. 1985a; Rompalo et al. 1986). When present, symptoms are usually mild and consist of mucoid anal discharge, mucus or pus-streaking of the stool, and perianal pain (Stamm et al. 1984). At sigmoidoscopy the rectal mucosa often appears normal, but in some patients there are features of a distal proctitis – mucosal oedema, loss of vascular pattern, friability and small erosions. The rectal histology is sometimes normal but there may be a non-specific proctitis consisting of a mild increase in the number of chronic inflammatory cells and polymorphs within the lamina propria (Quinn et al. 1981).

Lymphogranuloma venereum infection is associated with a more severe proctitis, usually with systemic features. The patient complains of ano-rectal pain, tenesmus, fever, diarrhoea, passage of blood per anum and anal discharge. There is a marked proctitis with anal ulceration, but the inflammatory changes seldom extend more proximally than 12 cm from the dentate line (Levine et al. 1980; Quinn et al. 1981). Occasionally the inflammation may be more localised, with the formation of an irregular ulcerated mass in the lower rectum that may produce profuse haemorrhage (Mindel 1983; Klotz et al. 1983). Inguinal lymph node involvement may be a feature of lesions of the anal canal and distal rectum (Mindel 1983; Klotz et al. 1983). Perianal abscesses, strictures and fistulae in ano may result from untreated LGV infection (Greaves 1963).

Histologically (Geller et al. 1980; Levine et al. 1980; Quinn et al. 1981), there is a dense infiltration of the lamina propria, extending into the submucosa, of lymphocytes, plasma cells, histiocytes and sometimes eosinophils. Occasional well-formed granulomas with giant cells and focal areas of acute inflammation are also found. Crypt abscesses may be a feature.

Pharyngeal Infection

Chlamydiae are isolated infrequently from the pharynges of homosexual men. In a group of 150 men who attended an STD clinic in Glasgow, UK, the organism was cultured from the pharynx of only two patients, both of whom had had fellatio with men with chlamydial urethritis (McMillan et al. 1981b). Experimental studies in chimpanzees suggest that, unlike urethral infection that requires low inocula, infection of the pharynx is only established with large inocula of chlamydiae (Jacobs et al. 1978). This finding may explain the relative rarity of pharyngeal infection in humans, even after fellatio with an infected partner (Bowie et al. 1977). Although *C. trachomatis* has been isolated from patients with pharyngitis (Schachter and Atwood 1975; Watanakunakorn and Levy 1983), there is little evidence to suggest that the organism is an important cause of pharyngitis. In a study of 95 young adults with pharyngitis, *C. trachomatis* was not isolated from any patient (Gerber et al. 1984).

Diagnosis

Laboratory diagnosis of chlamydial infection is discussed in detail in the monograph by Oriel and Ridgway (1982). The diagnosis of chlamydial proctitis is made by culture of the organism from rectal material. Although cotton wool-tipped applicator sticks, calcium alginate swabs and polyester sponges have been used for the collection of specimens, the author prefers to use a Jones–Dunlop urethral curette. Unfortunately, in a proportion of specimens isolation of chlamydiae may be impossible because of growth in the media of faecal commensals. Limited experience with direct immunofluorescence of smears suggests that this may be a useful technique for the examination of rectal material. Although standard staining of histological sections for microorganisms usually yields negative results, chlamydial antigen has been detected by immunofluorescence in the rectal mucosa from a patient with LGV proctitis (Klotz et al. 1983).

Serological tests (complement fixation and microimmunofluorescence) for serum antibodies against chlamydiae have proved to be useful in the diagnosis of LGV. High titres (>16) or significant changes in complement-fixing antibody titres in paired sera, or the detection of high titres (>512) of specific immunoglobulins, IgG or IgM, have been used as diagnostic criteria (Quinn et al. 1981). Schachter (1981), however, has noted that rectal infection with serovars of biovar *trachomatis* can induce antibodies that react with the LGV serovars. This finding and the genus-specific nature of the complement fixation test, limit the value of serology in the differentiation of rectal infection with LGV from oculogenital strains.

Although the complement fixation test is strongly positive and the microimmuno-fluorescent test shows the presence of high titred anti-chlamydia IgG in almost every case of rectal LGV (Schachter 1981), infection of HIV-infected men may not stimulate an antibody response. The author has studied two patients with persistent generalized lymphadenopathy who developed acute haemorrhagic proctitis associated with *C. trachomatis* serovar L2. Chlamydial antibodies were not detected in paired sera from either patient, who also failed to develop an antibody response to keyhole limpet haemocyanin, a potent immunogen.

The serological response to rectal infection with non-LGV immunovars is variable. When antibodies are detected by the microimmunofluorescence test, they are generally of low titre (Quinn et al. 1981). As 23% to 40% of "normal" adults have detectable serum antibody (Taylor-Robinson and Thomas 1980), the value of serological testing of single sera in diagnosis is very limited. Seroconversion or a significant change in antibody titre in paired sera, however, may give useful information.

Treatment

In vitro, chlamydiae are sensitive to the action of the tetracyclines and the macrolides erythromycin and rosaramicin. In clinical practice the tetracyclines have proved most useful in the treatment of urethral chlamydial infections (Oriel and Ridgway 1982). The drug, given as oxytetracycline, doxycycline or minocycline should be given for at least seven days.

Optimum treatment regimens for rectal chlamydial infection have not been reported, but published data (summarised in Table 2.9) suggest that the tetracyclines, used for at least two weeks, are satisfactory. In most cases, efficacy of treatment has been assessed only on the resolution of clinical features. Ideally, cultures should be obtained at an interval after completion of therapy.

Mycoplasma and *Ureaplasma* Infections

The genus *Mycoplasma* contains about 50 species, including at least five that can infect the genito-urinary tract and the genus *Ureaplasma* is represented by a single species, *Ureaplasma urealyticum*, that contains at least eight serotypes.

Mycoplasma hominis and *Ureaplasma urealyticum* are commonly cultured from rectal material from homosexual men. In one study, the isolation rates for both organisms were 34% (Munday et al. 1981a). There was no association between the presence of these organisms and clinical features of proctitis. With respect to *M. hominis*, a similar conclusion was reached by Quinn et al. (1983). They, however, isolated *U. urealyticum* more frequently from the rectums of men with proctitis than from those of control subjects.

Interestingly, although there was no significant difference in the isolation rate of *U. urealyticum* from their urethras the organism was isolated from the rectums of eight of 20 homosexual men, but from none of 34 heterosexuals. *Mycoplasma hominis* was detected in the rectums of three of 20 homosexual and one of 32 heterosexual men. *Mycoplasma genitalium* was cultured with equal frequency from the rectums of heterosexual and homosexual men, especially when the latter had non-gonococcal urethritis (Taylor-Robinson et al. 1985).

Table 2.9. Treatment of rectal chlamydial infections

Number of patients	Serotype of infecting organism	Treatment regimen	Method used for assessment of cure	Reference
2	—[a]	Tetracycline 500 mg q.i.d. for 2 weeks	Resolution of clinical features	Geller et al. (1980)
3	—[a]	Trimethoprim (160 mg)/sulpha-methoxazole (800 mg) b.d. for 4 weeks (2 cases). Trimethoprim/ sulphamethoxazole b.d. and tetracycline 500 mg q.i.d. for 3 weeks (1 case)	Resolution of clinical features	Levine et al. (1980)
14	L$_2$ (3 cases) Non-LGV (8 cases) Non-typed (3 cases)	Tetracycline 500 mg q.i.d. for 2 to 3 weeks	Resolution of clinical features	Quinn et al. (1981)
3	L2	Doxycycline 100 mg b.d. for 3 weeks Trimethoprim–sulphamethoxazole b.d. for 2 weeks followed by erythromycin 250 mg q.i.d. for 1 week Tetracycline 500 mg q.i.d. for 3 weeks	Resolution of clinical features and negative culture (2 cases)	Bolan et al. (1982)
1	L2	Tetracycline 500 mg q.i.d. for 3 weeks	Resolution of clinical features	Klotz et al. (1983)
1	—[a]	Oxytetracycline 500 mg q.i.d. for 3 weeks	Resolution of clinical features	Mindel (1983)
17	L2 (5 cases) Non-LGV (12 cases)	Tetracycline 500 mg q.i.d. for 1 to 2 weeks	Resolution of clinical features and negative post-treatment cultures	Stamm et al. (1984)
2	L2	Doxycycline 100 mg b.d. for 2 weeks	Resolution of clinical features	A. McMillan (unpublished work)

b.d. = twice daily; q.i.d. = four times per day.
[a]Diagnosis of LGV made on serological testing.

These data suggest that *U. urealyticum* and *M. hominis* are transmitted to the rectum during anal intercourse, but their role as rectal pathogens remains to be elucidated.

Rectal Spirochaetosis

Although for several decades spirochaetes had been identified in faeces from patients with diarrhoea, spirochaetes adherent to the microvillous surface of the columnar epithelial cells of the rectal mucosa were first noted by Harland and Lee (1967) in electron micrographs of a rectal biopsy from a 64 year old man with diarrhoea. Such colonisation is indicated in haematoxylin- and eosin-stained sections by the presence of a haematoxyphil zone, 3 μm wide, on the luminal surface of the cells (Fig. 2.1).

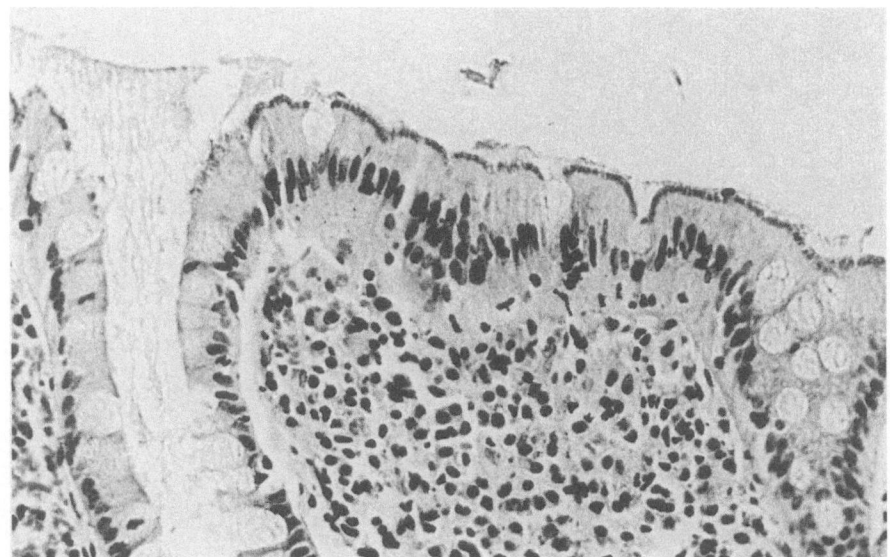

Fig. 2.1. Haematoxylin–eosin-stained section of rectal mucosa showing rectal spirochaetosis. ×320.

Although this condition is referred to as *intestinal spirochaetosis* this may be a misnomer. Certainly intestinal spirochaetosis has been noted in the appendices of patients with rectal colonisation, but this may not always be so, and the author prefers to refer to this condition as rectal spirochaetosis (RS).

In biopsies taken from patients attending general medical or surgical departments, the prevalence of RS is between 2% and 7% (Lee et al. 1971; Takeuchi et al. 1974; Gilmour unpublished data). McMillan and Lee (1981), however, observed the condition in 36 of 100 homosexual men who attended consecutively an STD clinic in Glasgow.

By electron microscopy, the spirochaetes associated with RS are 3–6 μm long and 0.2–0.4 μm wide, with tapering ends and two to six spirals. The cytoplasm is contained within a cytoplasmic membrane and between this and the cell wall is a space containing a group of four to six axial filaments that arise from terminal discs at each end and wind helically around the cytoplasmic cylinder, interdigitating in the middle (Takeuchi et al. 1974). The long axes of the spirochaetes lie parallel to the microvilli of the epithelial cells of the lumen of the rectum and the superficial portions of the crypts (Fig. 2.2). Goblet cells are not affected (Takeuchi et al. 1974). In one study (Takeuchi et al. 1974), a flagellated bacterium was also noted in biopsies from patients with RS. Other electron microscopy studies did not mention the presence of these organisms (Crucioli and Busuttil 1981; Cooper et al. 1986).

Until recently, difficulty was experienced in culturing the spirochaete associated with RS. Using a blood agar containing spectinomycin and incubating anaerobically, Tompkins et al. (1981) cultured large spirochaetes from three of 25 rectal swabs from homosexual men and from the faeces of one man. Smaller organisms with four axial filaments were cultured on a spectinomycin containing blood agar from patients with histologically proved RS (Hovind-Hougen et al. 1982) and as these spirochaetes differed from other species it was named *Brachyspira aalborgi*. There were, however,

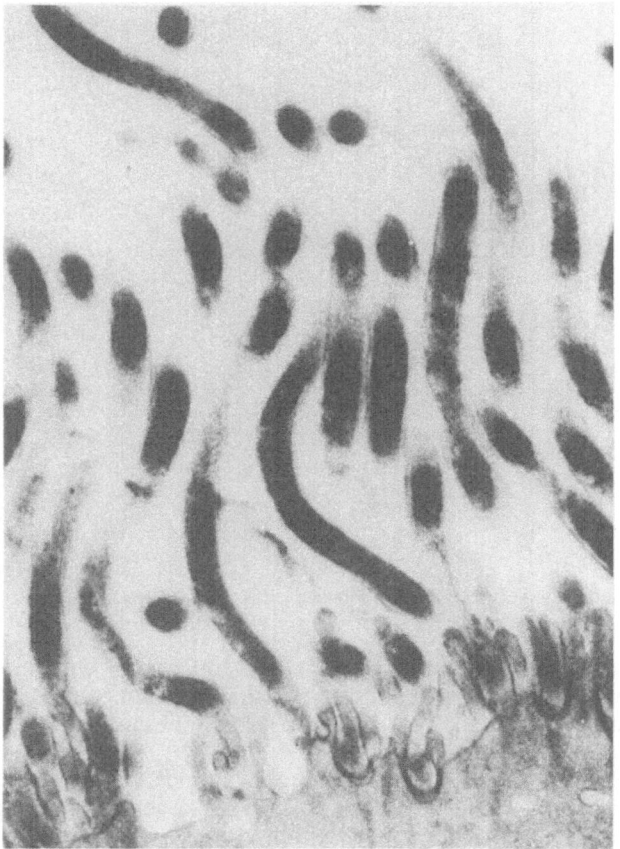

Fig. 2.2 Transmission electron micrograph of rectal mucosa showing spirochaetosis. ×42.

biochemical differences between the spirochaetes cultured from the rectums of seven of 34 homosexual men in West Yorkshire and *B. aalborgi* (Tompkins et al. 1986) and Cooper et al. (1986) noted that their isolates were less nutritionally fastidious, grew faster and had different growth temperature requirements from *B. aalborgi*. From the results of these studies, it seems probable that the rectum can be colonised by a heterogeneous group of spirochaetes.

The source of the rectal spirochaetes is uncertain, but their increased prevalence amongst homosexual men suggests sexual transmission. Characterisation of spirochaetes cultured from the oropharynx, urethra and subpreputial sac of sexual partners is however, needed before such a conclusion can be substantiated.

The pathogenicity of intestinal spirochaetes is controversial. Kaplan and Takeuchi (1979) cultured a spirochaete from rectal exudate from a homosexual man with a rectal discharge that resolved rapidly after treatment with penicillin. There was no histological evidence of RS, and other causes of his symptoms had not been completely excluded. Similar criticism may be made of the cases reported by Crucioli and Busuttil (1981), Douglas and Crucioli (1981), Tompkins et al. (1982) and Cotton et al. (1984), who described various intestinal symptoms in patients with RS.

Table 2.10. Prevalence of inflammatory changes in the rectal mucosa of 159 homosexual men with rectal spirochaetosis but in whom pathogenic bacteria, viruses and protozoa were not identified (A. McMillan and H. Gilmour, unpublished work)

Presence or absence of rectal spirochaetosis	Number of patients with biopsy gradings		Total
	A	B	
Present	34	17	51
Absent	85	23	108
Total	119	40	159

A, normal histology; B, increased number of chronic inflammatory cells in lamina propria.

In the absence of other intestinal infections or diseases, Neilsen et al. (1983) found no correlation between the presence of RS and intestinal symptoms in heterosexual patients. In a study of 159 homosexual men who attended the department of Genito-Urinary Medicine in Edinburgh, and in whom microbiological investigations showed no evidence of pathogenic bacterial, viral or protozoal infection, there was no correlation between RS and proctitis as assessed histologically (Table 2.10: A. McMillan and H. Gilmour, unpublished work). Cooper et al. (1986), however, noted that there was a generalised loss of microvilli from the epithelial cells of the rectal mucosa of homosexual men, particularly in those colonised by spirochaetes. The significance of this finding is uncertain, but the authors thought that the adherent organisms may exert a pathogenic effect through blockage of passive absorption without eliciting an inflammatory response.

Rectal spirochaetosis can persist for years in patients who have not received anti-microbial agents. Spontaneous loss of RS, however, has been noted (H. Gilmour, unpublished work).

All strains of spirochaetes isolated by Tompkins et al. (1986) were sensitive to metronidazole, tetracycline, fusidic acid and chloramphenicol, but were resistant to vancomycin and colistin. Some strains, of which some seemed to produce a β-lactamase, were resistant to penicillin.

Intestinal symptoms in some patients with RS have resolved after treatment with metronidazole (Douglas and Crucioli 1981; Crucioli and Busuttil 1981; Cotton et al. 1984). The author, however, feels that there is insufficient evidence to suggest that individuals with RS require treatment. In symptomatic patients, careful investigation for other causes of intestinal symptoms, including irritable bowel syndrome, should be made.

Shigellosis

On the basis of their biochemical characteristics, four species within the genus *Shigella* are recognised: *S. dysenteriae*, *S. flexneri*, *S. boydi* and *S. sonnei*. *Shigella dysenteriae* is subdivided, with 10 antigenically distinct types, *S. flexneri* into eight types that are antigenically interrelated and *S. boydi* into 15 types that are antigenically distinct.

Dritz and Back (1974) were the first to recognise the sexual transmission of *Shigella* spp.; they noted that 60% of 50 patients with enteritis due to *S. flexneri* in San Francisco were homosexual men. In a subsequent report from the same city (Dritz et al. 1977), it was shown that 70% of cases of shigellosis in 1976 occurred in men aged 20 to 39 years. Many of these men had had frequent orogenital and oro-anal sexual contact with male partners. An increased prevalence of *Shigella* infection amongst homosexual men who had not travelled abroad was also recorded in New York City (Drusin et al. 1976) and an outbreak of *S. flexneri* and *S. sonnei* dysentery in Seattle, King County, was reported by Bader et al. (1977).

Although many patients have very mild intestinal symptoms, others have more marked features of enteritis. The prepatent period is usually two to three days. Although the severity of the illness varies, *S. sonnei* dysentery is usually mild, the patient passing about ten loose stools during the first day. Thereafter, the frequency of passage of stools decreases until normal bowel habit is restored, generally within one week. A more severe illness with abdominal pain and tenderness, systemic features, and the passage of blood-stained mucus is more likely with *S. flexneri* than with *S. sonnei* infection. If sigmoidoscopy is undertaken in these patients, inflammatory changes are found within the rectum and extending proximally beyond the rectosigmoid junction. The histology is that of acute infective proctitis with no distinguishing features (Mandal et al. 1982).

The diagnosis is made by culture of faeces on selective media. Treatment is aimed at correcting dehydration and antimicrobial therapy is seldom indicated. It is generally accepted that three consecutive negative stool samples indicate resolution of the infection.

Salmonellosis

Salmonella typhi and *paratyphi* are usually acquired by ingestion of contaminated water, milk or foodstuff, direct infection from one individual to another being unusual.

Two homosexual men who had probably acquired typhoid fever from anilingus with symptomless carriers of *S. typhi*, however, were described by Dritz and Braff (1977).

Campylobacter Infections

Campylobacters are curved, Gram-negative bacilli with polar flagella. They are microaerophilic. Within the genus *Campylobacter* there are several species, but until recently only *C. fetus* was thought to be important in human disease. Classification within the species *fetus* is based on serotype and biotype, and three subspecies are recognised – *fetus*, *intestinalis*, and *jejuni* (Parker and Smith 1983). These subspecies are distinguishable by their ability to produce H_2S, glycine tolerance, growth at 25°C and 42°C and pathogenicity. *Campylobacter fetus* ssp. *fetus* rarely infects humans but can cause diarrhoea and systemic disease in the immunocompromised. *Campylobac-*

ter foetus ssp. *jejuni* is the organism most frequently isolated from the stools of individuals with campylobacter-associated diarrhoea. Two new species of *Campylobacter* that had been cultured from the rectums of homosexual men were described by Totten et al. (1985), *C. cinaedi* and *C. fennelliae*. These new species differ from *C. foetus* ssp. *jejuni* in their optimum growth temperature, biochemical reactivity and DNA homology. The pathogenicity of these two species is uncertain, but acute proctitis was a histological feature of all three infected patients studied by Quinn et al. (1984). These workers also identified a third strain of *Campylobacter*-like organism but, as its guanosine and cytosine content was higher than that of other species, its taxonomic status remains to be decided.

Most cases of *Campylobacter* diarrhoea are sporadic, the infection probably having been acquired from the ingestion of contaminated food, particularly poultry. Milk- and water-borne outbreaks, however, have been described. Although person to person spread is unusual, the infection has been acquired from faecally incontinent patients (Blaser et al. 1981), and hence homosexual men are at risk because of faecal contamination during sexual activity. Sporadic cases of *Campylobacter* enteritis in homosexuals have been reported (Carey and Wright 1979; Quinn et al. 1980), but the source of their infection had not been traced. Campylobacters were not recovered from the rectums of any of 50 homosexual men who attended a London STD clinic (Simmons and Tabaqchali 1979) and in a serological study (McMillan et al. 1984) the prevalence of IgG antibodies against *C. jejuni* in the sera of homosexual and heterosexual men who attended an STD clinic in Edinburgh was similar: 10% and 9% of 187 and 169 men, respectively.

In a study from Seattle, *C. jejuni* was isolated from 10 of 158 men with, and two of 75 homosexual men without, intestinal symptoms; campylobacters were not isolated from 150 heterosexual men and women (Quinn et al. 1984). *Campylobacter cinaedi* and *C. fennelliae* were identified in rectal material from 23 and six of 201 symptomatic homosexual men, respectively, and 13 and none of 155 symptomless men, respectively (Totten et al. 1985). In these studies, there was a significant association between *Campylobacter* infection and the practice of anilingus.

Campylobacter enteritis is a self-limiting diarrhoeal illness, usually of less than one week's duration. Cramping, generalised abdominal pain is a salient feature that may precede the onset of diarrhoea by 24 hours. The stools are loose and may contain mucus and blood. Fever, anorexia, malaise, myalgia, arthralgia and headache are common features (Blaser et al. 1979). Sigmoidoscopy during the acute phase of the illness shows a marked proctitis, the inflammatory changes being seen to extend proximally beyond the rectosigmoid junction. The histology is that of an acute infective proctitis (Lambert et al. 1979). When radiological examination has been undertaken, a pancolitis has been noted (Lambert et al. 1979); this investigation, however, is seldom indicated in the investigation of infective proctocolitis.

In some patients the course of the illness is prolonged, symptoms persisting for up to three weeks.

The median duration of excretion of *Campylobacter* is two to three weeks, but some patients continue to pass the organism in their faeces for up to three months (Blaser and Reller 1981). The diagnosis is made by culture of the organism on selective media containing blood or other animal fluid and incubated at a temperature of 42°C except when *C. fennelliae* and *C. cinaedi* are suspected, when the temperature should be 37°C.

Agglutinin and immunofluorescence methods have been used for the detection of serum antibody against campylobacters (Watson et al. 1979; Blaser et al. 1979).

High-titred antibody, a significant change in titre in paired sera, or the detection of specific IgM are considered to be diagnostic of recent infection.

As the enteritis associated with *Campylobacter* infection is usually of less than one week's duration, antimicrobial treatment is seldom indicated. In more persistent infections, or when there is a risk of transmission to others, for example by homosexual anal intercourse, erythromycin stearate in an oral dosage of 500 mg twice daily may be given.

Miscellaneous Infections

Urinary Tract Infections

In the absence of underlying disorders such as congenital abnormalities and calculi, urinary tract infection with eubacteria is uncommon in men under the age of 50 years. Although a case of haemorrhagic cystitis and epididymitis associated with *Salmonella enteritidis* was reported by Greene et al. (1982), until recently an increased prevalence of such infection in homosexual men was not described. Significant bacteriuria, however, was found in 14 of 280 sexually active young men with acute urinary symptoms who attended a STD clinic in Seattle; 12 of these men had had homosexual contact (Barnes et al. 1986). In this series, *Escherichia coli* was the most common infecting organism and it was suggested that the urethra may become colonized with faecal *E. coli*. Indeed non-gonococcal urethritis was found in more than half of the bacteriuric men, and in some cases its clinical features preceded those of the bladder infection.

Different results were reported by Wilson et al. (1986) working at the Middlesex Hospital, London. Urine samples from 200 homosexual and 205 heterosexual men were cultured for eubacteria. *Escherichia coli* was isolated from six patients, three of whom were homosexual. The authors concluded that the prevalence of urinary tract infection among homosexual men was similar to that in the heterosexual population.

These differences in prevalence of infection may reflect differences in patient selection and sexual practices, details of which were not given. In the author's experience, eubacterial infection of the urinary tract is uncommon in homosexual men, but he has noted symptomatic infection with *E. coli* in two men who introduced catheters into the bladder as part of sexual foreplay.

Culture of a midstream specimen of urine for eubacteria seems reasonable in homosexual men with clinical features of urethritis and cystitis but in whom *N. gonorrhoeae* and *C. trachomatis* have not been identified. Antimicrobial therapy depends on the results of the sensitivity testing of the isolate.

Streptococcal Infections

Streptococcal infection of the skin of the penis in a homosexual man has been described by Drusin et al. (1975). The lesion is erythematous, exudes pus and crusting may develop. In immunocompromised individuals, inguinal lymphadenitis associated with β-haemolytic streptococci may occur (Ho and Murata 1984).

References

Ansink-Schipper MC, Klingeren Bv, Huikeshoven MH, Woudstra RK, Dessens-Kroon M, Wijngaarden LJv (1984) Epidemiology of PPNG infections in the Netherlands: analysis by auxanographic typing and plasmid identification. Br J Vener Dis 60: 141–146

Austin TW, Lent B, Pattison FLM (1978) Gonorrhea in homosexual men. Can Med Assoc J 119: 731–732

Backman M, Ruden A-KM, Bygdeman SM, Jonsson A, Ringertz O, Sandstrom EG (1985) Gonococcal serovar distribution in Stockholm with special reference to multiple infections and infected partners. Acta Pathol Microbiol Scand Sect B 93: 225–232

Bader M, Pedersen AHB, Williams R, Spearman J, Anderson H (1977) Venereal transmission of shigellosis in Seattle-King County. Sex Transm Dis 4: 89–91

Barnes RC, Daifuku R, Roddy RE, Stamm WE (1986) Urinary tract infection in sexually active homosexual men. Lancet i: 171–173

Bisaillon JG, Turgeon P, Dubreuil D, Beaudet R, Sylvestre M, Ashton FE (1984) In vitro inhibition of growth of *Neisseria gonorrhoeae* by *Neisseria meningitidis* isolated from the pharynx of homosexual men. Sex Transm Dis II: 296–299

Blaser MJ, Reller BL (1981) *Campylobacter* enteritis. N Engl J Med 305: 1444–1452

Blaser MJ, Berkowitz ID, LaForce FM, Cravens J, Reller B, Wang W-L L (1979) *Campylobacter* enteritis: clinical and epidemiologic features. Ann Intern Med 91: 179–185

Blaser MJ, Waldman RJ, Barrett T, Erlandson AL (1981) Outbreaks of *campylobacter* enteritis in two extended families: evidence for person-to-person transmission. J Pediatr 98: 254–257

Bolan RK, Sands M, Schachter J, Miner RC, Drew WL (1982) Lymphogranuloma venereum and acute ulcerative proctitis. Am J Med 72: 703–706

Bowie WR, Alexander ER, Holmes KK (1977) Chlamydial pharyngitis? Sex Transm Dis 4: 140–141

Bowie WR, Alexander ER, Holmes KK (1978) Etiologies of postgonococcal urethritis in homosexual and heterosexual men: roles of *Chlamydia trachomatis* and *Ureaplasma urealyticum*. Sex Transm Dis 5: 151–154

Bro-Jorgensen A, Jensen T (1973) Gonococcal pharyngeal infections. Br J Vener Dis 49: 491–499

Bygdeman S (1981) Gonorrhoea in men with homosexual contacts. Serogroups of isolated gonococcal strains related to antibiotic susceptibility, site of infection and symptoms. Br J Vener Dis 57: 320–324

Bygdeman S (1987) Polyclonal and monoclonal antibodies applied to the epidemiology of gonococcal infection. In: Young H, McMillan A (eds) Diagnosis of sexually transmissible diseases with monoclonal and polyclonal antibodies. Marcel Dekker Inc., New York, Chap 4, pp 117–165

Bygdeman S, Danielsson D, Sandstron E (1981) Serological classification of *Neisseria gonorrhoeae* by coagglutination. A study of serological patterns in two geographical areas of Sweden. Acta Dermatol Venereol [Suppl] (Stockh) 61: 423–428

Bygdeman S, Backman M, Danielsson D, Norgren M (1982) Genetic linkage between serogroup specificity and antibiotic resistance in *Neisseria gonorrhoeae*. Acta Pathol Microbiol Scand Sect B 90: 243–250

Bygdeman SM, Mårdh PA, Sandstrom EG (1984) Susceptibility of *Neisseria gonorrhoeae* to rifumpicin and thiamphenicol: correlation with protein I antigenic determinants. Sex Transm Dis 11: 366–370

Carey PB, Wright EP (1979) *Campylobacter jejuni* in a male homosexual Br J Vener Dis 55: 380

Carlson BL, Haley MS (1984) Single-site infection with *Neisseria gonorrhoeae* in homosexual men. Sex Transm Dis 11: 312–313

Carlson BL, Fiumara NJ, Kelly JR, McCormack WM (1980) Isolation of *Neisseria meningitidis* from anogenital specimens from homosexual men. Sex Transm Dis 7: 71–73

Centers for Disease Control (1982) Sexually transmitted diseases treatment guidelines. MMWR 31(25)

Cooper P, Cotton DWK, Hudson MJ, Kirkham N, Willmott FEW (1986) Rectal spirochaetosis in homosexual men: characterisation of the organism and pathophysiology. Genitourin Med 62: 47–52

Cotton DWK, Kirkham N, Hicks DA (1984) Rectal spirochaetosis. Br J Vener Dis 60: 106–109

Crawford G, Knapp JS, Hale J et al. (1977) Asymptomatic gonorrhea in men: caused by gonococci with unique nutritional requirements. Science 196: 1352–1353

Crucioli V, Busuttil A (1981) Human intestinal spirochaetosis. Scand J Gastroenterol 16 (Suppl 70): 177–178

Deheragoda P (1977) Diagnosis of rectal gonorrhoea by blind anorectal swab compared with direct vision swab taken via a proctoscope. Br J Vener Dis 53: 311–313

DiCaprio JM, Reynolds J, Frank G et al. (1978) Ampicillin therapy for pharyngeal gonorrhea. JAMA 239: 1631–1633

Douglas JG, Crucioli V (1981) Spirochaetosis: a remediable cause of diarrhoea and rectal bleeding? Br Med J 283: 1362

Dritz SK, Back AF (1974) *Shigella* enteritis venereally transmitted. N Engl J Med 291: 1194

Dritz SK, Braff EH (1977) Sexually transmitted typhoid fever. N Engl J Med 296: 1359

Dritz SK, Ainswort TE, Garrard WF et al. (1977) Patterns of sexually transmitted enteric diseases in a city. Lancet ii: 3–4

Drusin LM, Wilkes BM, Gingrich RD (1975) Streptococcal pyoderma of the penis following fellatio. Br J Vener Dis 51: 61–62

Drusin LM, Genvert G, Topf-Olstein B, Levy-Zombeck E (1976) Shigellosis. Another sexually transmitted disease? Br J Vener Dis 52: 348–350

Evans DTP, Crooks AJR, Jones C, Holman RA, Price SW (1986) Treatment of uncomplicated gonorrhoea with single dose aztreonam. Genitourin Med 62: 318–320

Fagan D (1985) Comparison of *Neisseria gonorrhoeae* isolates from homosexual and heterosexual men. Genitourin Med 61: 363–366

Faur YC, Wilson ME, May PS (1981) Isolation of *N. meningitidis* from patients in a gonorrhea screening program: a four-year survey in New York City. Am J Public Health 71: 53–58

Fiumara NJ (1978) The treatment of gonococcal proctitis. An evaluation of 173 patients treated with 4 g of spectinomycin. JAMA 239: 735–737

Fiumara NJ (1979) Pharyngeal infection with *Neisseria gonorrhoeae*. Sex Transm Dis 6: 264–266

Fluker JL, Boulton Hewitt A (1970) Kanamycin in the treatment of rectal gonorrhoea. Br J Vener Dis 46: 454–456

Fluker JL, Deherogoda P, Platt DJ, Gerken A (1980) Rectal gonorrhoea in male homosexuals. Presentation and therapy. Br J Vener Dis 56: 397–399

Geller SA, Zimmerman MJ, Cohen A (1980) Rectal biopsy in early lymphogranuloma venereum proctitis. Am J Gastroenterol 74: 433–435

Gerber MA, Ryan RW, Tilton RC, Watson JE (1984) Role of *Chlamydia trachomatis* in acute pharyngitis in young adults. J Clin Microbiol 20: 993–994

Givan KF, Jaeger R (1986) Auxotypes of *Neisseria gonorrhoeae* isolated from heterosexual men, homosexual men and heterosexual women. Sex Transm Dis 13: 19–23

Givan KF, Thomas BW, Johnston AG (1977) Isolation of *Neisseria meningitidis* from the urethra, cervix and anal canal: further observations. Br J Vener Dis 53: 109–112

Goh BT, Varia KB, Ayllife PP, Lim FKS (1985) Diagnosis of gonorrhea by Gram-stained smears and cultures in men and women: role of the urethral smear. Sex Transm Dis 12: 135–139

Goldmeier D, Darougar S (1977) Isolation of *Chlamydia trachomatis* from throat and rectum of homosexual men. Br J Vener Dis 53: 184–185

Graudal C, Bollerup AC, Lange K, Seier K, Lind I (1985) The outcome of single-dose cefuroxime treatment in patients with pharyngeal gonorrhea. Sex Transm Dis 12: 49–51

Greaves AB (1963) The frequency of lymphogranuloma venereum in persons with perirectal abcesses, fistulae in ano or both. Bull WHO 29: 797–801

Greaves WL, Kraus SJ, McCormack WM, Biddle JW, Zaidi A, Fiumara NJ, Guinan ME (1983) Cefoxitin vs penicillin in the treatment of uncomplicated gonorrhea. Sex Transm Dis 10: 53–55

Greene JB, Adler M, Holzman RS (1982) *Salmonella enteritidis* genitourinary tract infection in a homosexual man. J Urol 128: 1046–1048

Hallqvist L, Lindgren S (1975) Gonorrhoea of the throat at a venereological clinic. Incidence and results of treatment. Br J Vener Dis 51: 395–397

Handsfield HH (1983) Treatment of uncomplicated gonorrhea in men with single-dose moxalactam. Sex Transm Dis 10: 191–194

Handsfield HH, Holmes KK (1981) Treatment of uncomplicated gonorrhea with cefotaxime. Sex Transm Dis 8: 187–191

Handsfield HH, Murphy VL (1983) Comparative study of ceftriaxone and spectinomycin for treatment of uncomplicated gonorrhoea in men. Lancet ii: 67–70

Handsfield HH, Knapp JS, Diehr PK, Holmes KK (1980) Correlation of auxotype and penicillin susceptibility of *Neisseria gonorrhoeae* with sexual preference and clinical manifestations of gonorrhea. Sex Transm Dis 7: 1–5

Harland WA, Lee FD (1967) Intestinal spirochaetosis. Br Med J 3: 718–719

Ho DD, Murata GH (1984) Streptococcal lymphadenitis in homosexual men with lymphadenopathy. Am J Med 77: 151–153

Holmes KK, Counts GW, Beaty HN (1971) Disseminated gonococcal infection. Ann Intern Med 74: 979–993

Hovind-Hougen K, Anderson AB, Neilsen RH et al. (1982) Intestinal spirochaetosis: morphological characterisation and cultivation of the spirochete *Brachyspira aalborgi* gen. nov. sp. nov. J Clin Microbiol 16: 1127–1136

Jacobs NF, Arum ES, Kraus SJ (1978) Experimental infection of the chimpanzee urethra and pharynx with *Chlamydia trachomatis*. Sex Transm Dis 5: 132–136

Jaffe HW, Biddle JW, Johnson SR, Wiesner PJ (1981) Infections due to penicillinase producing *Neisseria gonorrhoeae* in the United States: 1976–1980. J Infect Dis 144: 191–197

Janda WM, Bohnhoff M, Morello JA, Lerner SA (1980) Prevalence and site pathogen studies of *Neisseria meningitidis* and *N. gonorrhoeae* in homosexual men. JAMA 244: 2060–2064

Janda WM, Morello JA, Lerner SA, Bohnhoff M (1983) Characteristics of pathogenic *Neisseria* spp. isolated from homosexual men. J Clin Microbiol 17: 85–91

John J, Jefferiss FJG (1973) Treatment of anorectal gonorrhoea with ampicillin. Br J Vener Dis 49: 362–363

Judson FN (1983) Fear of AIDS and gonorrhoea rates in homosexual men. Lancet ii: 159–160

Judson FN, Ehret JM, Eickhoff TC (1978) Anogenital infection with *Neisseria meningitidis* in homosexual men. J Infect Dis 137: 458–463

Judson FN, Ehret JM, Hunter Handsfield H (1985) Comparative study of ceftriaxone and spectinomycin for treatment of pharyngeal and anorectal gonorrhea. JAMA 253: 1417–1419

Kaplan LR, Takeuchi A (1979) Purulent rectal discharge associated with a non-treponemal spirochete. JAMA 241: 52–53

Karney WW, Pedersen AHB, Nelson M et al. (1977) Spectinomycin versus tetracycline for the treatment of gonorrhea. N Engl J Med 296: 889–894

Klotz SA, Drutz DJ, Tam MR, Reed KH (1983) Hemorrhagic proctitis due to lymphogranuloma venereum serogroup L2. Diagnosis by fluorescent monoclonal antibody. N Engl J Med 308: 1563–1565

Lambert ME, Schofield PF, Ironside AG, Mandal BK (1979) *Campylobacter* colitis. Br Med J 1: 857–859

Lawrence AG, Shanson DC (1985) Single dose oral amoxicillin 3 g with either 125 mg or 250 mg clavulanic acid to treat uncomplicated anogenital gonorrhoea. Genitourin Med 61: 168–171

Lebedeff DA, Hochman EB (1980) Rectal gonorrhea in men: diagnosis and treatment. Ann Intern Med 92: 463–466

Lee FD, Kraszewski A, Gordon J, Howie JGR, McSeveney D, Harland WA (1971) Intestinal spirochaetosis. Gut 12: 126–133

Levine JS, Smith PD, Brugge WR (1980) Chronic proctitis in male homosexuals due to lymphogranuloma venereum. Gastroenterology 79: 563–565

Lindberg M, Ringerty O, Sandstrom E (1982) Treatment of pharyngeal gonorrhoea due to β-lactamase producing gonococci. Br J Vener Dis 58: 101–104

Mandal BK, Schofield PF, Morson BC (1982) A clinicopathological study of acute colitis: the dilemma of transient colitis syndrome. Scand J Gastroenterol 17: 865–869

McMillan A, Lee FD (1981) Sigmoidoscipic and microscopic appearance of the rectal mucosa in homosexual men. Gut 22: 1035–1041

McMillan A, Young H (1978) Gonorrhea in the homosexual man: frequency of infection by culture site. Sex Transm Dis 5: 146–150

McMillan A, Sommerville RG, McKie PMK (1981) Chlamydial infection in homosexual men: frequency of isolation of *Chlamydia trachomatis* from the urethra, ano-rectum and pharynx. Br J Vener Dis 57: 47–49

McMillan A, McNeillage G, Gilmour H, Lee FD (1983) Histology of rectal gonorrhoea in men with a note on anorectal infection with *Neisseria meningitidis*. J Clin Pathol 36: 511–514

McMillan A, McNeillage GJC, Watson KC (1984) The prevalence of antibodies reactive with *Campylobacter jejuni* in the serum of homosexual men. J Infect 9: 63–68

Metzger AL (1970) Gonococcal arthritis complicating gonococcal pharyngitis. Ann Intern Med 73: 267–269

Miller MA, Millikin P, Griffin PS et al. (1979) *Neisseria meningitidis* urethritis: a case report. JAMA 242: 1656–1657

Mindel A (1983) Lymphogranuloma venereum of the rectum in a homosexual man. Case Report. Br J Vener Dis 59: 196–197

Morrison GD, Evans AJ, Haskins HW, Lewis N McL, Seale GH, Mayall E, Mullinger BM (1980) Cefuroxime compared with penicillin for the treatment of gonorrhea. Sex Transm Dis 7: 188–190

Morse SA, Lysko PG, McFarland L, Knapp JS, Sandstrom E, Critchlow C, Holmes KK (1982) Gonococcal strains from homosexual men have outer membranes with reduced permeability to hydrophobic molecules. Infect Immun 37: 432–438

Munday PE, Dawson, SG, Johnson AP et al. (1981a) A microbiological study of non-gonococcal proctitis in passive male homosexuals. Postgrad Med J 57: 705–711

Munday PE, Thomas BJ, Johnson AP, Altman DG, Taylor-Robinson D (1981b) Clinical and microbiological study of non-gonococcal urethritis with particular reference to non-chlamydial disease. Br J Vener Dis 57: 327–333

Munday PE, Carder JM, Taylor-Robinson D (1985a) Chlamydial proctitis? Genitourin Med 61: 376–378

Munday PE, Bingham JS, Ison CA, Erdman YJ, Harris JRW, Easmon CSF (1985b) Treatment of gonorrhea with clavulanate-potentiated amoxicillin (Augmentin). Sex Transm Dis 12: 163–165

Nielsen RH, Orholm M, Pederson JO, Hovind-Hougen K, Teglbjaerg PS, Thaysen EH (1983) Colorectal spirochaetosis: clinical significance of the infestation. Gastroenterology 85: 62–67

Odegaard K, Gundersen T (1973) Gonococcal pharyngeal infection. Br J Vener Dis 49: 350–352

Oriel JD, Ridgway GL (1982) Genital infections by *Chlamydia trachomatis*. Edward Arnold, London

Oriel JD, Reeve P, Wright JT, Owen J (1976) Chlamydial infection of the male urethra. Br J Vener Dis 52: 46–51

Osborne NG, Grubin L (1979) Colonization of the pharynx with *Neisseria gonorrhoeae*; experience in a clinic for sexually transmitted diseases. Sex Transm Dis 6: 253–256

Owen RL, Hill JL (1972) Rectal and pharyngeal gonorrhea in homosexual men. JAMA 220: 1315–1318

Parker MT, Smith G (1983) *Vibrio, Aeromonas, Plesiomonas, Campylobacter* and *Spirillum*. In Wilson G, Miles A, Parker MT (eds) Principles of Bacteriology, Virology and Immunity, vol 2. Edward Arnold, London, pp 137–155

Porter IA, Rutherford HW (1977) Treatment of uncomplicated gonorrhoea with spectinomycin hydrochloride (Trobicin). Br J Vener Dis 53: 115–117

Poulsen A, Ullman S (1985) AIDS induced decline of the incidence of syphilis in Denmark. Acta Dermatol Venereol 65: 567–569

Price JD, Fluker JL (1975) Amoxicillin in the treatment of gonorrhoea. Br J Vener Dis 51: 398–400

Price JD, Fluker JL, Giles AJH (1977) Oral talampicillin in the treatment of gonorrhoea. Br J Vener Dis 53: 113–114

Quinn TC, Corey L, Chaffee RG, Schuffler MD, Holmes KK (1980) *Campylobacter* proctitis in a homosexual man. Ann Intern Med 93: 458–459

Quinn TC, Goodell SE, Mkrtichian PA-C et al. (1981) *Chlamydia trachomatis* proctitis. N Engl J Med 305: 195–200

Quinn TC, Stamm WE, Goodell SE et al. (1983) The polymicrobial origin of intestinal infections in homosexual men. N Engl J Med 309: 576–582

Quinn TC, Goodell SE, Fennell C et al. (1984) Infections with *Campylobacter jejuni* and *Campylobacter* like organisms in homosexual men. Ann Intern Med 101: 187–192

Reid KG, Young H (1984) Serogrouping *Neisseria gonorrhoeae*: correlation of coagglutination serogroup WII with homosexually acquired infection. Br J Vener Dis 60: 302–305

Rompalo AM, Price CB, Roberts PL, Stamm WE (1986) Potential value of rectal-screening cultures for *Chlamydia trachomatis* in homosexual men. J Infect Dis 153: 888–892

Ruden A-KM, Werner YK, Ringertz O, Bygdeman SM, Backman M, Sandstrom EG (1985) Use of gonococcal W serogrouping in the evaluation of a clinical trial of rosoxacin. Sex Transm Dis 12: 19–24

Salit IE, Frasch CE (1982) Seroepidemiologic aspects of *Neisseria meningitidis* in homosexual men. Can Med Assoc J 126: 38–41

Sands M (1980) Treatment of anorectal gonorrhea infections in men. JAMA 243: 1143–1144

Schachter J (1981) Confirmatory serodiagnosis of lymphogranuloma venereum proctitis may yield false positive results due to other chlamydial infections of the rectum. Sex Transm Dis 8: 26–27

Schachter J, Atwood G (1975) Chlamydial pharyngitis? J Am Vener Dis Assoc 2: 12

Scott J, Stone AH (1966) Some observations on the diagnosis of rectal gonorrhoea in both sexes using a selective culture medium. Br J Vener Dis 42: 103–106

Silva AH de, Bashi SAQ, Roy RB (1984) Treatment of uncomplicated anogenital gonorrhoea with a single oral dose of 3 g amoxicillin combined with 250 mg clavulanic acid. Br J Vener Dis 60: 132–133

Simmons PD, Tabaqchali S (1979) *Campylobacter* species in male homosexuals. Br J Vener Dis 55: 66

Spencer RC, Smith T, Talbot MD (1984) Ceftizoxime in the treatment of uncomplicated gonorrhoea. Br J Vener Dis 60: 90–91

Stamm WE (1984) Proctitis due to *Chlamydia trachomitis*. In Ma P, Armstrong D (eds) The acquired immune deficiency syndrome in infections of homosexual men. Yorke Medical Books, New York, pp 40–47

Stamm WE, Quinn TC, Mkrtichian EE, Wang SP, Schuffler MD, Holmes KK (1982) *Chlamydra trachomitis* proctitis. In: Mårdh PA, Holmes KK, Oriel JD, Piot P, Schachter J (eds) Chlamydial infections. Elsevier Biomedical Press, Amsterdam, pp 111–114

Stamm WE, Koutsky LA, Benedetti JK, Jourden JL, Brunham RC, Holmes KK (1984) *Chlamydia trachomatis* urethral infections in men. Prevalence, risk factors and clinical manifestations. Ann Intern Med 100: 47–51

Stolz E, Schuller J (1974) Gonococcal oro- and nasopharyngeal infection. Br J Vener Dis 50: 104–108

Stolz E, Ong L, van Joost T, Michel MF (1984) Treatment of uncomplicated urogenital rectal and oropharyngeal gonorrhoea with intramuscular cefotaxime 1.0 g or cefuroxime 1.5 g. J Antimicrob Chemother 14 [Suppl B]: 295–299

Takeuchi A, Jervis HR, Nakazawa H, Robinson DM (1974) Spiral shaped organisms on the surface colonic epithelium of the monkey and man. Am J Clin Nutr 27: 1287–1296

Taylor PK, Seth AD (1975) Ampicillin plus probenecid compared with procaine penicillin plus probenecid in the treatment of gonorrhoea. Br J Vener Dis 51:183–187

Taylor-Robinson D, Thomas BJ (1980) The role of *Chlamydia trachomatis* in genital tract and associated diseases. J Clin Pathol 33: 205–233

Taylor-Robinson D, Furr PM, Hanna NF (1985) Microbiological and serological study of non-gonococcal urethritis with special reference to *Mycoplasma genitalium*. Genitourin Med 61: 319–324

Thin RNT (1974) Penicillin treatment of gonorrhoea in Edinburgh. Br J Vener Dis 50: 57–60

Tompkins DS, Waugh MA, Cooke EM (1981) Isolation of intestinal spirochaetes from homosexuals. J Clin Pathol 34: 1385–1387

Tompkins DS, Cooke EM, MacDonald RC, Abbott CR (1982) Spirochaetosis: a remediable cause of diarrhoea and rectal bleeding? Br Med J 284: 52

Tompkins DS, Soulkes SJ, Goodwin PGR, West AP (1986) Isolation and characterisation of intestinal spirochaetes. J Clin Pathol 39: 535–541

Totten PA, Fennell CL, Tenover FC et al. (1985) *Campylobacter cinaedi* (sp. nov.) and *Campylobacter fennelliae* (sp. nov.): two new *Campylobacter* species associated with enteric disease in homosexual men. J Infect Dis 151: 131–139

Wallin J, Siegel MS (1979) Pharyngeal *Neisseria gonorrhoeae*: coloniser or pathogen. Br Med J 2: 1462–1463

Wanas TM, Williams PEO (1986) Oral cefuroxime axetil compared with oral ampicillin in treating acute uncomplicated gonorrhoea. Genitourin Med 62: 221–223

Watanakunakorn C, Levy DH (1983) Pharyngitis and urethritis due to *Chlamydia trachomatis*. J Infect Dis 147: 364

Watson KC, Kerr EJC, McFadzean SM (1979) Serology of human campylobacter infections. J Infect 1: 151–152

Waugh MA (1970) Trimethoprim-sulphamethoxazole (Septrin) in the treatment of rectal gonorrhoea. Br J Vener Dis 47: 34–35

Weller IVD, Hindley DJ, Adler MW, Meldrum JT (1984) Gonorrhoea in homosexual men and media coverage of the acquired immune deficiency syndrome in London 1982–3. Br Med J 289: 1041

Wiesner PJ, Tronca E, Bonin P, Pedersen AHB, Holmes KK (1973) Clinical spectrum of pharyngeal gonococcal infection. N Engl J Med 288: 181–185

Wilson APR, Tovey SJ, Adler MW, Gruneberg RN (1986) Prevalence of urinary tract infection in homosexual and heterosexual man. Genitourin Med 62: 189–190

Chapter 3

Viral Infections

A. Mindel

Introduction

A large number of viral infections may be sexually transmitted in homosexual men. These include hepatitis A, hepatitis B, warts and human immunodeficiency virus infection, all of which are considered elsewhere in this book, as well as genital herpes, cytomegalovirus infection, and molluscum contagiosum, which will be considered in this chapter.

Genital Herpes

Genital herpes has been recognised for many centuries. However, it is only in recent years that the disease has achieved more prominence amongst both the medical profession and the lay public. There are several reasons for this, including the increasing willingness of people to attend at sexually transmitted disease (STD) clinics, the possible association of herpes with cervical cancer, the recognition of the devastating infection that can occur in the neonate and the immunocompromised, the availability of technology to study the molecular biology of herpes and finally the recent introduction of an effective and safe drug for its treatment.

Virology

Herpes simplex virus (HSV) is a DNA virus 130–200 nm in size and icosahedral in shape. The virus consists of four distinct morphological elements: an electron-opaque core, a capsid enclosing the core, an electron-dense tegument and the envelope (Nahmias and Roizman 1973a).

The core contains the linear double-stranded viral DNA with a molecular weight of approximately 100×10^6. The DNA is large enough to encode 60 to 70 or more gene products. The DNA sequence homology between HSV1 and HSV2 is about 50%, and homologous gene sequences are distributed over the entire genome map (Roizman and Frenkel 1973).

The diameter of the capsid is 100 nm and it consists of 162 capsomers arranged in the form of an icosahedron. Each capsomere resembles a hexagonal prism with a hollow duct running parallel to the long axis (Wildy et al. 1960).

The envelope forms a loose impermeable coat around the virus particle. It contains a lipid bilayer embedded in which are five or six viral glycoproteins, which are important in the attachment of the virus to the host cell and the subsequent penetration into the cell. These glycoproteins are designated: gB, which is required for infectivity; gC which binds to the C3b complement component; gD, which may be required for infectivity and is the most potent neutralising antibody inducer; and gE, which binds to the Fc portion of IgG (Spear 1984). The gG glycoprotein shows little cross-reactivity between HSV1 and HSV2 and may be a useful candidate for a type-specific serological assay (Lee et al. 1985, 1986).

Biosynthesis of the Virus

The virus enters the cell by fusion of the surface membrane with the plasma membrane of the cell. Little is known about the mechanism of attachment of viral glycoprotein to the cell membrane or whether specific receptors are involved. The capsid migrates to the nuclear pore and releases its DNA or a nucleoprotein complex into the nucleus. DNA synthesis and transcription of RNA occur in the nucleus (Rapp 1984).

The synthesis of virus-specific gene products appears to be regulated by three classes of HSV genes, designated alpha, beta and gamma (Nahmias and Roizman 1973a; Roizman and Furlong 1974). The initial genes (alpha) are expressed earliest and do not require any prior viral protein synthesis. The exact function of the alpha proteins is unknown, but they are required for the production of the beta proteins, which are in turn responsible for the production of proteins and enzymes (including DNA polymerase and thymidine kinase) and for induction of the gamma class of HSV genes (Leiden et al. 1976; Powel and Purifoy 1977). The expression of these genes is dependent on viral DNA replication and most of the structural components of the virus particles are gamma proteins (Wagner 1985; Roizman and Furlong 1974).

The DNA synthetic pathway depends initially on the phosphorylation of deoxyribonucleosides by viral thymidine kinase. The resulting deoxyribonucleoside monophosphate is further phosphorylated into the di- and triphosphate derivates by cellular enzymes. The deoxyribonucleoside triphosphate is finally converted to DNA by DNA polymerase.

The DNA is then processed into capsids, which are enveloped by the nuclear membrane and leave the cell by the way of the endoplasmic reticulum.

Epidemiology

Since the human being is the only host for HSV1 and HSV2, the infected person is the sole reservoir. Infection occurs when susceptible individuals come into contact with infected secretions during close personal contact. Such contact can be genital, oral or anal. As these viruses are very unstable, this appears to be the only important method of transmission. The incubation period is two to 14 days for both viruses, with initial infection occurring at the site of inoculation (Rawls and Campione-Piccardo 1981). The most severe attacks occur in non-immune individuals (that is those with no previous exposure to HSV). Latency is a particular property of these viruses, with periodic reactivation in the dermatomal distribution of the site originally infected (Spruance et al. 1977; August et al. 1979). Virus excretion associated with such reactivations constitutes the major source of infection.

Most oral infections are caused by HSV1. Primary mouth infections are rare in children under six months of age, presumably due to passively acquired maternal antibodies (Adam 1982). However, subsequent infections in young children are common. Acquisition of oral herpes appears to be related largely to age and socioeconomic status (Nahmias et al. 1970; Wentworth and Alexander 1971).

Antibodies to HSV persist for many years after initial exposure to the virus, and seroepidemiological studies are therefore an accurate reflection of past exposure. Unfortunately the interpretation of serological tests has two major limitations: firstly, there is considerable cross-reactivity between HSV1 and HSV2 and, secondly, the two viral types can cause either oral or genital infections.

The majority of genital and anal infections are caused by HSV2, which is spread by sexual intercourse with a partner who has genital sores at the time (Mindel and Adler 1984) or who is excreting the virus asymptomatically. HSV1 genital infections are contracted by oral–genital contact with a person who has active cold sores; however, patients with inapparent infections are also important. A recent study of 69 contacts of patients with first-episode genital herpes found that only 26% knew that they had herpes prior to transmission. However, history and examination revealed that 65% had evidence of oral or genital herpes (Mertz et al. 1985).

Trends

Genital herpes was first reported from STD clinics in the UK in 1972 when 4501 were seen. The number of cases has increased each year and by 1984 the number reported was 19 869 (Communicable Disease Surveillance Centre 1985) (Fig. 3.1). In the 1970s the male to female ratio was approximately 3 : 1; however, since 1981, the gap has narrowed. Although the sexual orientation of males is not reported, a recent study from London showed that 11% of patients with herpes in 1972 and 20% in 1982 were homosexual (Hindley and Adler 1985).

The apparent increase in herpes infection may be due partly to increased publicity and the inclusion of both primary and recurrent cases in clinic returns (Hindley and Adler 1985). On the other hand the size of the increase suggests that a considerable part of it is real.

The situation in other countries is less clear. In the USA no accurate national data are available; however, estimates in the popular press put the number of sufferers at between 2 million and 20 million.

The Centers for Disease Communication have estimated that there are between 300 000 and 500 000 new cases of herpes every year in the USA, on the basis of a survey of ten STD clinics (Centers for Disease Communication 1982). However, the validity of this figure is questionable as the majority of patients attending such clinics are of lower socioeconomic status and do not represent a reasonable cross-section of the population.

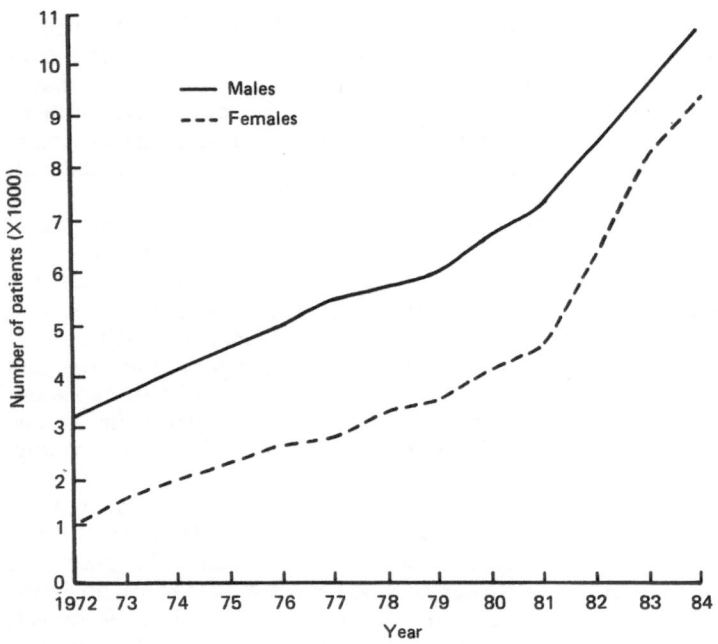

Fig. 3.1. Genital herpes reported from STD clinics in the United Kingdom 1972–1984 (from Mindel et al. 1984).

In Scandinavia, herpes is also common. A survey in a Swedish STD clinic found that herpes could be isolated from 8.4% of the cervices of all female attenders (Jeansson and Molin, 1974) and the situation in other Western countries is probably similar. Virtually nothing is known about the prevalence in most Third World countries.

Natural History

Latency and Reactivation

It is a characteristic of all herpesviruses that they may persist in the host, probably in a non-infectious form, with the possibility of subsequent reactivation.

Following a first attack of herpes (oral, genital or anal), HSV rapidly finds virus receptors that are present on sensory nerve terminals (Vahlne et al. 1978). Studies by Lycke and co-workers (1984) suggest that the axon is penetrated as part of the virus replicative process. Having penetrated the axon cylinder, the virus is inaccessible to the immune system and is translocated centripetally by axoplasmic movement (Kristensson 1978). Recent studies in mice suggest that HSV becomes immunologically privileged as early as 48 hours after inoculation (Simmons and Nash 1984).

Once latency is established the infection appears to reside in a dormant state in the cell bodies of neurones in sensory and autonomic ganglia within the central nervous system (Puga et al. 1978).

Reactivation of infective virus may lead to clinical recurrences at the periphery, but need not inevitably follow it (Wildy et al. 1982). The reasons for reactivation and its exact mechanisms have not been fully elucidated. It has long been known that patients with herpes often have a relapse when subjected to non-specific stimuli such as sunburn, fever or menstruation (Wildy 1985; Guinan et al. 1981).

The role of the immune system has not been fully elucidated (Wildy 1985); however, conditions associated with depressed cell-mediated immunity often present with severe herpetic relapses (Hill 1985; Stevens 1978). This situation is commonly seen in patients with haematological malignancies (Muller et al. 1972), malnutrition (Becker et al. 1968), renal allografts (Montgomerie et al. 1969) and the acquired immune deficiency syndrome (AIDS) (Siegal et al. 1981).

Viral type is also important in determining the likelihood and frequency of subsequent recurrences. HSV2 infection recurs more frequently in patients than does HSV1 (Reeves et al. 1981).

Clinical Features – First Episode

Primary infection can involve the penis, the perianal area, the anus or the rectum. In the last of these, the disease is often very severe and prolonged, whereas penile herpes is usually a minor complaint. The illness may be caused by either HSV1 or HSV2 and the clinical features are indistinguishable. In contrast the patients' antibody status at the time of the first genital episode is an important determinant of disease severity. Those with antibodies have a less severe illness, of shorter duration, than do those without (Corey et al. 1983a).

Symptoms of perianal and anal herpes include rectal discharge and pain, tenesmus and flu-like symptoms. After a few days, perianal lesions may be noticed. On examination the patient is febrile, the inguinal lymph nodes are often enlarged and tender, perianal vesicles or ulcers are often seen. Proctoscopy is excruciatingly painful and the rectal mucosa is erythematous, friable and covered in mucopus. On occasions, localised ulceration may be seen. Inflammation is confined to the lower 10 cm of the rectum (Goldmeier 1980; Goodell et al. 1983). The illness is often protracted, with systemic symptoms lasting up to two weeks and the perianal and anal lesions taking up to four weeks to heal (Samarasinghe et al. 1979; Waugh 1976).

Goodell and co-workers studied the clinical features of 23 patients with herpes proctitis and compared them with 79 patients with non-herpes proctitis. Significantly more patients with HSV proctitis had ano-rectal pain, tenesmus, constipation, pruritis, perianal lesions, inguinal lymphadenopathy and fever compared with those whose proctitis was not due to HSV.

Penile herpes by contrast is usually relatively mild and short lived. Presenting symptoms include pain in the penis or groin, dysuria and flu-like symptoms. Examination reveals bilateral large, tender, inguinal lymph nodes, fever, and numerous lesions on the penis (Corey et al. 1983a). The lesions commence as vesicles that ulcerate and then heal. The ulcers often have an erythematous halo and a greyish white base. On the skin, ulcers heal by crusting, but on the mucous membranes no crusting occurs (Davis and Keeney 1981). Lesions can occur on any penile site; however, the glans, the coronal sulcus and the foreskin are the commonest. Symptoms settle in a few days and healing is complete by two weeks (Corey et al. 1983a).

Complication of First Attack Genital Herpes

A number of complications have been described during or following the first attack of genital herpes, including dissemination to sites distant from the genitalia, meningitis, transverse myelitis and autonomic nervous system dysfunction, urinary difficulties or retention, necrotising balanitis, urethral stricture, suppurative lymphangitis, and secondary bacterial or fungal infection.

Extragenital involvement may occur from primary inoculation at sites such as the fingers or pharynx, from haematogenous spread during the viraemic phase of the illness, or from autoinoculation to any mucocutaneous site (Corey et al. 1983a). Corey and co-workers (1983a) reported that the commonest sites of extragenital involvement were the fingers and areas adjacent to the genitalia, suggesting that lesion arose from autoinoculation rather than viraemia.

Meningitis occurs in 13% of men with primary genital herpes (Corey et al. 1983a). The clinical features include fever, headache, photophobia, neck-stiffness and a positive Kernig's sign. The cerebrospinal fluid shows a slight increase in both protein and lymphocytes. The condition, in common with most viral meningitides, resolves within a few days without residual neurological sequelae (Skoldenberg et al. 1975; Meyer et al. 1960; Corey et al. 1983a).

Transverse myelitis and autonomic nervous system dysfunction both occur with genital herpes (Klastersky 1977; Caplan et al. 1977; Craig and Nahmias 1973) and appear to be particularly common in homosexual men with herpetic proctitis (Oates and Greenhouse 1978; Samarsinghe et al. 1979). Signs and symptoms of the autonomic nervous system dysfunction include: hyperaesthesia or anaesthesia in the perineum, thighs or buttocks, with decreased sensation over the sacral dermatomes; difficulty with urination and defaecation; poor rectal and perianal sphincter tone; an enlarged bladder and an absent bulbocavernosis reflex (Riehle and Williams 1979; Oates and Greenhouse 1978; Jacome and Yanez 1980; Goldmeier et al. 1975; Samarasinghe et al. 1979).

Urinary problems commonly occur in patients with first attack genital herpes because of severe pain associated with urethral or peri-urethral lesions (Nahmias and Roizman 1973b) or because of the transverse myelitis or autonomic nervous system dysfunction. The problem is self-limiting; however, in severe cases catheterisation may be necessary (Corey et al. 1983a).

Rare complications of penile herpes include urethral stricture, necrotising balanitis (Powers et al. 1982) and suppurative lymphangitis of the dorsum of the penis (Tottie 1943).

Recurrent Herpes

Recurrences are usually of shorter duration and lesser severity than primary infections. They usually occur at the site of the original primary and consist of a single or a small group of lesions, often at a single anatomical site usually on the external genitals, buttocks or perianal area. Systemic symptoms are uncommon; however, many patients complain of warning or prodromal symptoms consisting of burning or pain in the genitals or a neuralgia-type pain radiating into the buttocks, groin or the back of the thighs. The prodrome starts a few hours or even a few days before lesions are visible. The neuralgia pain often persists for several days. The entire episode last seven to ten days (Guinan et al. 1981; Corey et al. 1983a). Although the symptoms

Table 3.1. Differences between primary and recurrent genital herpes

	First attack	Recurrences
Pain/dysuria	Severe	Absent–mild
No. of lesions	Multiple	Few
Anatomical sites	Multiple	Usually single
Constitutional symptoms	Common	Rare
Prodromal symptoms	Unusual	Common
Duration of lesions (days)	7–28	2–10
Duration of pain (days)	5–20	1–7
Duration of viral shedding (days)	5–25	1–10
Radiculomyelopathy	Up to 50%	—
Meningitis	± 10%	—

and signs of recurrent genital herpes are often trivial, the recurrent nature of the illness may give rise to severe psychosexual dysfunction (Adler and Mindel 1983). The differences between primary and recurrent herpes are summarised in Table 3.1.

Diagnosis

Clinical Diagnosis

The clinical diagnosis of penile herpes often poses no problem, particularly where vesicles are present. Occasionally, however, other causes of genital ulceration will need to be considered (Table 3.2). The differential diagnosis of penile ulceration is enormous; however, several clinical features may point to the diagnosis of herpes. These include the presence of vesicles or multiple painful ulcers with tender bilateral inguinal lymphadenopathy. Other conditions that commonly cause multiple painful genital ulcers include herpes zoster involving the genital area, Behçet's syndrome and chancroid. Tender, enlarged lymph nodes are a feature of chancroid and herpes zoster but not Behçet's syndrome.

The clinical diagnosis of anal herpes is often more difficult. There are several causes of painful proctitis, including gonorrhoea and *Chlamydia*. Goodell and co-workers in 1983 studied the clinical features of men with herpetic proctitis and compared them with those with proctitis due to other conditions. Ano-rectal pain and tenesmus occurred in all the patients with herpes but also in 77% of those with non-herpetic proctitis. In contrast the presence of constipation, pruritis, inguinal lymphadenopathy, perianal lesions, fever and neurological symptoms were useful pointers to the diagnosis.

Laboratory Diagnosis

The cultivation of HSV from infected secretions or tissues is the most widely used and sensitive technique in the diagnosis of genital herpes. Other techniques that may be used include the detection of viral particles or viral antigen in infected tissues, and the use of serological tests.

Viral culture is the most reliable method of confirming the diagnosis. Material containing the infected cells (e.g. vesicle fluid or cells from the base of ulcers or inflamed mucosa) should be transported to the laboratory in a suitable transport medium, such as Hanks balanced salt solution with antibiotics, or veal infusion broth.

Table 3.2. Causes of genital ulceration

Infections

Genital herpes	– primary
	recurrent
Syphilis	– primary chancre
	– secondary mucous patches
	– gummatous

Candidosis
Trichomoniasis
Scabies
Chancroid
Lymphogranuloma venereum
Granuloma inguinale
Pyogenic infection
Folliculitis
Herpes zoster

Non-infective causes
Trauma – physical (e.g. sexual intercourse or masturbation)
       – chemical (e.g. antiseptics, caustic agents etc.)
Reiters syndrome – circinate balanitis
Behçet's syndrome
Leukoplakia
Lichen sclerosis et atrophicus
Neoplasia (carcinoma penis/vulva)
Crohn's disease
Drug reactions (e.g. Stevens–Johnson syndrome)

HSV may be cultured in a variety of cell lines, including human diploid fibroblasts lines such as human embryonic tonsil (HET), embryonic lung, embryonic kidney (HEK) or human amnion cells, and various animal cells including rabbit kidney, African green monkey kidney (VERO) or baby hamster kidney (GBK) cells (Rawls 1979; McSwiggan et al. 1975).

HSV produces a typical cytopathic effect, commencing as rounding of cells, which eventually become swollen and refractile, and ultimately die and detach from the surface of the culture plate. Cytopathic effect can be seen as early as 12 hours or as late as seven days after inoculation.

There are a number of biological differences between HSV1 and HSV2 that can be used for typing isolates. Firstly, HSV1 grows poorly, whereas HSV2 grows well, in primary chick embryo fibroblast culture (Figuero and Rawls 1969). Secondly, HSV1 produces small pocks and HSV2 large pocks, on chorionic membranes of embryonated hens eggs (Nahmias and Dowdle 1968). Thirdly, low concentration of deoxyribonucleosides such as thymidine, markedly inhibit the production of CPE in cell lines infected with HSV2 whereas those infected with HSV1 show no such inhibition (Kelman et al. 1975).

A number of immunological techniques have been employed for the differentiation of the two viral types. These tests all rely on the fact that the reaction is more intense with the homologous than with the heterologous antiserum, and all are limited by the considerable cross-reactivity between HSV1 and HSV2. Tests that have been used include an indirect immunoperoxidase test (Benjamin 1977), a direct immunofluorescence test (Nahmias et al. 1971) and an enzyme-linked immunosorbent assay (Vestergaard and Grauballe 1979). These tests can be used for rapid diagnosis to replace or supplement viral culture. However, they all have the major drawback of being less sensitive than HSV cultures. The introduction of monoclonal antibodies to HSV subtypes has improved the sensitivity and specificity of these tests (Chan 1983).

Other rapid diagnostic tests include staining lesion scrapings to identify multi-nucleated giant cells (Tzanck test; Naib 1966), and electron microscopy from lesion scrapings. This latter technique does not differentiate between HSV and other herpesviruses.

The recent introduction of restriction enzyme technology has opened the way not only for unambiguous differentiation of HSV1 and HSV2 but also for mapping of individual strains (Lonsdale 1979). This technique relies on cleavage of double-stranded DNA by certain restriction endonucleases to produce fragments, the electrophoretic pattern of which is highly characteristic for that DNA.

The serological diagnosis of genital herpes is complicated firstly by the diversity of assays, secondly by the cross-reactivity between HSV1 and HSV2, and thirdly because recurrent genital HSV does not produce a rise in complement-fixing (CF) or neutralising antibodies.

Serological assays include CF neutralisation, passive haemagglutination, indirect haemagglutination, immune adherence, ELISA and radioimmunoassay (RIA). Measurement of IgM antibodies has been described (Doerr et al. 1976) and is a consistent finding in primary genital or orolabial herpes but not with recurrent disease.

A newly described immunodot enzymatic assay using purified glycoprotein prepared with monoclonal antibodies to type-specific HSV2 and HSV1 glycoprotein G (designated gG2 and gG1) appears to eliminate the problem of cross reactivity between HSV1 and HSV2 and is a potentially exciting new development (Lee et al. 1985, 1986).

Treatment

Numerous and diverse therapies have been tried for the treatment of genital herpes. Belsey and Adler in 1978 found that 16 different therapies were being used routinely in STD clinics in the UK, and Corey from the other side of the Atlantic listed 23 treatments (Corey et al. 1981).

Table 3.3 lists the different treatments that have been used to treat genital herpes. The vast majority either show no efficacy or have not been evaluated in controlled clinical trials. The treatment of genital herpes involves several strategies. Firstly, treatment of the initial attack, secondly treatment of recurrences, and finally suppression or prevention of recurrences. The major aims of the treatment in primary herpes are to decrease the time to healing, the duration and severity of symptoms, and the duration of viral shedding, and to prevent development of complications (Adler and Mindel 1985).

The only drug that has been shown in a series of clinical trials to fulfil these aims is acyclovir (Mindel et al. 1982; Corey et al. 1982, 1983b; Nilsen et al. 1982; Bryson et al. 1983; Mertz et al. 1984a; Thin et al. 1983). The antiviral activity depends on its conversion to a triphosphate derivative, which acts as a substrate for DNA polymerase and ultimately in viral DNA chain termination. The specificity and safety of acyclovir depends on its initial specific phosphorylation by viral thymidine kinase (Elion 1982; Bridgen and Whiteman 1983). This drug is particularly useful in the treatment of primary herpetic proctitis (A.M. Rompalo et al., unpublished work, presented at 6th Internat Meeting Soc STD Res, 1985). In this situation, either intravenous acyclovir (5 mg/kg eight hourly, by slow infusion for five to ten days) or oral acyclovir (200 mg five times a day for five days) should be used. The earlier treatment is started the better the results. Treatment started when the lesions have been present for longer than six days is of questionable benefit. Unfortunately acyclovir does not prevent the development of subsequent recurrences. This is not surprising, as treatment is usually commenced only 10–14 days after exposure, by which time latency is well established and the virus immunologically privileged. As penile herpes is less severe, the use of acyclovir in this situation is of limited value. In addition to specific antiviral therapy, symptomatic treatment is also useful. Many patients require mild analgesia and some find that frequent bathing with dilute saline is soothing and promotes healing.

Table 3.3. Treatments evaluated in controlled trials in patients with genital herpes

Drug	Route	Results	Comments
Idoxuridine	Topical	Minor reduction in viral shedding	Of limited value Efficacy seen only with higher dosage
Adenine arabinoside	Topical	Ineffective	—
2-Deoxy-D-glucose	Topical	?Some reduction in viral shedding and time to healing	Only study, highly criticised
Ribavrin	O	? Decreased severity	Unproven
Acyclovir	IV O Topical }	Decreased duration of viral shedding, symptoms, and lesions	Consistent results with numerous studies
Topical surfactants Ether	Topical	Ineffective	Toxic
Nonoxynol 9	Topical	Ineffective	Toxic
Immune modulators Levamisole	O	Ineffective	Some toxicity
Inosine pranobex	O	Limited efficacy	Trials poorly conducted and reported
Transfer factor	O	Ineffective	
Interferon	i.m.	Limited efficacy	Unproven
Photodynamic inactivation Proflavine	Topical	Ineffective	Enhances oncogenic potential of HSV

O, oral; i.v., intravenous; i.m., intramuscular.

The treatment of recurrent infection is more problematic. The use of oral or topical acyclovir to treat each recurrence is of questionable clinical benefit. The drug significantly reduces the duration of viral shedding and the time to healing; however, the actual reduction in healing time is only in the region of one to one and a half days. The effect of symptoms is variable (Nilsen et al. 1982; Salo et al. 1983; Reichman et al. 1983, 1984; Corey et al. 1982). A more useful approach is the use of prophylactic acyclovir, where the drug is taken continuously to prevent the development of recurrences (Mindel et al. 1984; Douglas et al. 1984; Straus et al. 1984; Fiddian et al. 1983). This form of therapy has proved highly effective (Fig.3.2); however, a number of questions remain unanswered. What is the ideal dosage? Is the drug safe in the long term? How long should treatment continue? Who should be treated? Further studies will be required to answer these questions.

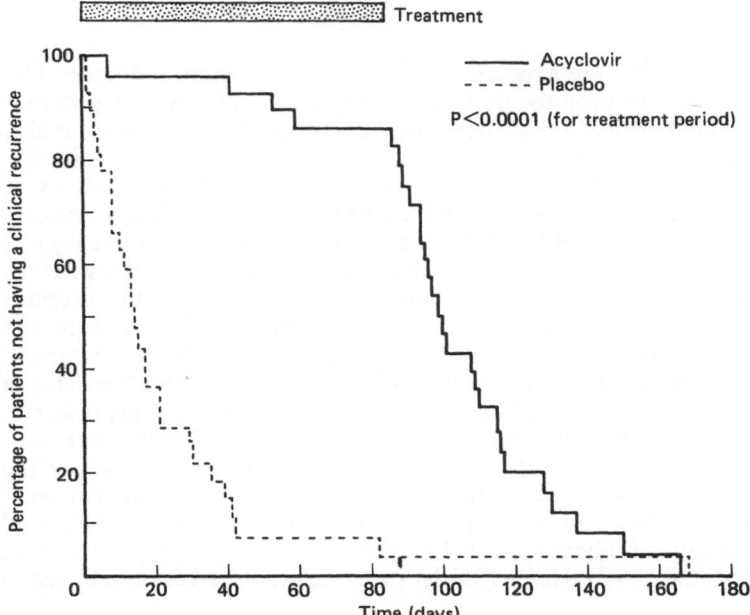

Fig. 3.2. Suppressive acyclovir treatment in patients with frequently recurring genital herpes – results of randomised double-blind placebo controlled trial. Only 14% of acyclovir recipients recurred during treatment compared with 98% of those receiving placebo. The recurrences in the acyclovir recipients were minor and transient (from Mindel et al. 1984).

Advice and Counselling

Genital herpes may be a very upsetting and stressful illness. Luby and Gillespie (1981) described a sequence of psychological events following the first attack of herpes.

1. Initial shock and emotional numbing
2. Search for a cure
3. Sense of isolation and loneliness
4. Anger directed at the person who was the source of the infection
5. Fear concerning the sexual consequences
6. Feelings of ugliness and contamination – the "leper effect"
7. Depression
8. Reactivation in rare cases of underlying psychopathology

Considerable patience, skill and sensitivity are often required to deal with these problems. A study by Manne and Sandler in 1984 has shown that patients can be helped by providing social support and by training them to think positively rather than concentrating on the negative aspects of the disease.

Prevention

The majority of people who are exposed to HSV develop subclinical disease. The fact that the infection can be acquired from an asymptomatic individual suggests that health education is unlikely to decrease transmission. Antiviral chemotherapy is also likely to have only a minimal impact.

The only realistic method of prevention is by vaccination. For a vaccine to work, it must be able to prevent the establishment of latency or reactivation and transmission of infection. (Corey and Spear 1986). Animal experiments suggest that latency can be prevented by the use of live or inactivated as well as purified-subunit vaccines (Allen and Rapp 1982; Lasky et al. 1984). Concern about the safety of live vaccines has been raised. HSV can transform cell lines in vitro, can cause malignant disease in animals and has been implicated as the causation of cervical and vulval carcinoma.

Subunit vaccines appear to offer the best hope. Animal experiments have shown that various components of the membrane glycoprotein offer protection (Lory et al. 1984). The vaccines are immunogenic in humans (Mertz et al. 1984b); however, further studies will be required to determine which HSV antigens (if any) are the most effective in inducing protective immunity. Two subunit vaccines are currently under evaluation in humans. One is a purified glycoprotein subunit type 2, said to contain no viral DNA (Hilleman et al. 1981); the other is a polypeptide subunit of HSV1 (Skinner et al. 1982). Neither of the two vaccines have yet been tested in controlled trials.

Herpes and Human Immunodeficiency Virus Infection

Patients infected with HIV infection often have particular problems with herpes. Recurrences of oral penile or anal herpes may become more severe and prolonged; however, more seriously, perianal and anal lesions may become very extensive and persistent (Siegel et al. 1981). The onset may be insidious, with periods of partial healing, but ultimately the ulcerated area persists and extends. These lesions respond to intravenous acyclovir (5–10 mg/kg eight hourly by slow intravenous infusion for five to ten days). Many HIV antibody patients may require continuous prophylactic acyclovir (see Chap. 8).

Cytomegalovirus

Introduction

Infection with cytomegalovirus (CMV) is widespread, with the majority of infections being acquired in childhood. Adult infections may be sexually transmitted and are common in homosexual men. In recent years, this virus has assumed new importance as a result of the HIV epidemic. In immunocompetent individuals, the disease is insignificant; however, in patients with AIDS the disease may be life threatening.

Epidemiology

CMV can be isolated from virtually all body fluids including saliva, vaginal and cervical secretions, and semen. Infection with CMV occurs through close interpersonal contact. Antibody studies have shown that in many poorer parts of the world the majority of people (95%–100%) are infected between the ages of three and five years (Krech 1973). In the UK, North America and Europe, 50%–60% of the population will eventually acquire the infection, a third in childhood and the rest between 15 and 35 years of age (Stern and Elek 1965). Many people acquiring the infection after the age of 15 may get their infection via the sexual route. CMV acquired from the mother during pregnancy is one of the most important causes of neonatal viral infections. The spectrum of neonatal illness includes hepatosplenomegaly, microcephaly, choroidoretinitis, mental retardation and deafness.

The evidence for the sexual spread of CMV is circumstantial. Firstly, CMV can be isolated from semen (Lang and Kummer 1972) as well as from cervical and vaginal secretions (Jordan et al. 1973). Secondly, CMV can be isolated more frequently from the cervices of females attending STD clinics than from those attending gynaeocology, family planning or well women clinics (Embil et al. 1985), and finally Chreitein and colleagues (1977) reported on the sexual transmission of two male cases of CMV mononucleosis from the same female.

The homosexual life-style appears to carry a very high risk of acquiring CMV. Several studies have shown the vast majority (71%–94%) of homosexual and bisexual men attending STD clinics have CMV antibodies (Drew et al. 1981; Mindel and Sutherland 1984; Coutinho et al. 1984) and that the percentage with antibodies is significantly higher than in randomly selected heterosexuals. Homosexuality was shown to outweigh the other known risk factors associated with the acquisition of antibody, namely social class, age and nationality (Mindel and Sutherland 1984).

The extraordinarily high prevalence of this infection in homosexual men possibly reflects the greater degree of sexual experimentation and the frequency of partner change in this group of patients.

Clinical Features and Diagnosis

The majority of adults with primary CMV infection remain asymptomatic. The few that do have symptoms develop malaise, fever, arthralgias and myalgias 30–40 days after exposure (Evans 1978; Cohen and Corey 1985). Examination reveals a pyrexia in virtually all patients and hepatosplenomegaly in about half. Unlike infectious mononucleosis, exudative tonsillitis and lymphadenopathy are unusual. A small minority of patients will develop hepatitis, bronchial pneumonia, polyneuritis, myocarditis or pericarditis (Lamb and Stern 1966; Carlstrom et al. 1968; Klemola et al. 1972; Schmitz and Enders 1977; Dietz 1981; Wilson et al. 1972). Virus can be recovered from polymorphonuclear leucocytes for the first two weeks of the illness and for up to three months from the throat, urine and semen. The virus can be grown in cell culture but it only produces its typical cytopathic effect two to three weeks after inoculation. The specific antibody response may take several weeks to develop. Atypical lymphocytes are seen in up to 40% of patients and many patients will have abnormal liver function tests, particularly a raised alkaline phosphatase. Using a monoclonal antibody to early virus-coded proteins, Griffiths and co-workers (1984) have devised a indirect immunofluorescence test that can detect early antigen fluorescent foci 24 hours after inoculation into cell cultures.

Cytomegalovirus in AIDS

Immunocompromised patients often develop severe disseminated infection with CMV. These infections may involve any organ; however, common sites involved include the lungs, the bowel and the retina. The clinical features of disseminated CMV in AIDS are discussed in Chap. 8.

Treatment

Infections in immunocompetent adults are self-limiting and do not require therapy. In immunocompromised patients with life-threatening infections, a number of new antiviral agents show some promise; however, none of these has been tested in controlled clinical trials. Two of these are phosphonoformate and 9-(2-hydroxy-1-(hydroxymethyl)ethoxymethy)guanine. Uncontrolled studies suggest that the latter agent may be particularly useful in halting the progression of CMV pneumonitis and retinitis. These agents are discussed in more detail later (see Chap. 8).

Molluscum Contagiosum

Molluscum contagiosum is a benign disease caused by an as yet unassigned pox virus. Epidemiological surveys suggest that infection occurs during two periods of life, in childhood and in young adults (Brown 1984). In the latter group, circumstantial evidence supports the view that transmission is associated with sexual intercourse. The evidence includes the frequent history of other sexually transmitted diseases in these patients (Cobbold and MacDonald 1970; Brown and Weinberger 1974; Wilkin 1977), the identification of lesions in sexual partners (Gudgel 1954; Jacobs 1970; Brown and Weinberger 1974; Wilkin 1977) and the anatomical distribution of the lesions.

The incubation period is between one and six months (Brown et al. 1982). The typical lesions consist of a papule with an unbilicated centre. Lesions vary in size from 2 to 5 ml. In sexually transmitted infections, lesions are distributed on the genital region, the inner thighs, buttocks and lower abdominal walls.

The condition is self-limiting; however, treatment may prevent further progression from autoinoculation. A number of forms of treatment have been used in this condition, but none of these has been tested in controlled clinical studies. Amongst the therapies that are used are caustic agents (e.g. silver nitrate or carbolic acids) and cryotherapy applied to the unbilicated centre of the lesion.

References

Adam E (1982) Herpes simplex virus infections. In: Glaser R, Gotlieb-Stematsky T (eds) Human herpes
virus infections: clinical aspects. Marcel Dekker, New York, pp 1–47

Adler MW, Mindel A (1983) Genital herpes – hype or hope (Leader). Br Med J 286: 1767–1768

Adler MW, Mindel A (1985) Genital infection: antiviral chemotherapy and interferon. Br Med Bull 41:
361–366

Allen WP, Rapp F (1982) Concept review of genital herpes vaccines. J Infect Dis 145: 413–420

August MJ, Nordlund JJ, Hsiung GD (1979) Persistence of herpes simplex virus type 1 and 2 in infected
individuals. Arch Dermatol 115: 309–310

Becker WB, Kipps A, McKenzie D (1968) Disseminated herpes simplex virus infection. Am J Dis Child
115: 1–8

Belsey EM, Adler MW (1978) Current approaches to the diagnosis of herpes genitalis. Br J Vener Dis 54:
115–117

Benjamin DR (1977) Use of immunoperoxidase for rapid diagnosis of mucocutaneous herpes simplex
virus infection. J Clin Microbiol 6: 511–573

Bridgen DM, Whiteman P (1983) The mechanism of action, pharmacokinetics and toxicity of acyclovir: a
review. J Infect 6: 3–9

Brown SJ (1984) Molluscum contagiosum. In: Holmes KK, Mardh PA, Sparling PF, Weisner PJ (eds)
Sexually transmitted diseases. McGraw-Hill Book Company, New York, pp 507–512

Brown SJ, Weinberger J (1974) Molluscum contagiosum: sexually transmitted disease in 17 cases. J Am
Vener Dis Assoc 1: 35–36

Brown SJ, Nalley JF, Kraus SJ (1982) Molluscum contagiosum. Sex Trans Dis 8: 227–234

Bryson YJ, Dillon M, Lovett M et al. (1983) Treatment of first episode genital herpes virus infection with
oral acyclovir. A randomised double blind controlled trial in normal subjects. N Engl J Med 308:
916–921

Caplan LR, Kleeman FJ, Berg S (1977) Urinary retention probably secondary to herpes genitalis. N Engl
J Med 297: 920

Carlstrom G, Alder J, Belfrage S et al. (1968) Acquired cytomegalovirus infection. Br Med J 2: 521–525

Centers for Disease Communication (1982) Genital herpes infections, United States 1966–1979. MMWR
31: 137–139

Chan WL (1983) Protective immunisation of mice with specific HSV-1 glycoproteins. Immunology 49:
343–352

Chreitein JH, McGinnis CG, Miller A (1977) Venereal causes of cytomegalovirus mononucleosis. JAMA
238: 1644–1645

Cobbold RJC, MacDonald A (1970) Molluscum contagiosum as a sexually transmitted disease. Prac-
titioner 204: 416–419

Cohen JI, Corey GR (1985) Cytomegalovirus infections in the normal host. Medicine (Baltimore) 64:
100–114

Communicable Disease Surveillance Centre and the Academic Department of Genito-Urinary Medicine
(1985) Sexually transmitted diseases 1983. Br Med J 291: 528–530

Corey L, Spear PG (1986) Infection with herpes simplex virus (First of two parts). N Engl J Med 314:
686–691

Corey L, Holmes KK, Benedetti J, Critchlow C (1981) Clinical course of genital herpes: implication for
therapeutic trial. In: Nahmias AJ, Dowdle WR, Schinazi RF (eds) The human herpesviruses an interdisciplinary approach, Elsevier, New York, pp 496–502

Corey L, Nahmias AJ, Guinan ME, Benedetti JK, Critchlow CW, Holmes KK (1982) A trial of topical
acyclovir in genital herpes virus infections. N Engl J Med 306: 1313–1319

Corey L, Adams HG, Brown ZP, Holmes KK (1983a) Genital herpes simplex virus infections: clinical
manifestations, course and complications. Ann Intern Med 96: 958–972

Corey L, Fife KH, Benedetti JK et al. (1983b) Intravenous acyclovir for the treatment of primary genital
herpes. Ann Intern Med 98: 914–921

Coutinho RA, Wertheim-van Dillen P, Albricht-van Lent P et al. (1984) Infection with cytomegalovirus
in homosexual men. Br Med J 60: 249–252

Craig C, Nahmias AJ (1973) Different patterns of neurologic involvement with herpes simplex virus types
1 and 2: isolation of herpes simplex virus type 2 from the Buffy coat of 2 adults with meningitis. J Infect
Dis 127: 365–372

Davis LG, Keeney RF (1981) Genital herpes simplex virus infection: clinical course and attempted
therapy. Am J Hosp Pharmacol 38: 825–829

Dietz AJ (1981) Cytomegalovirus infections with carditis, hepatitis and anemia. Postgrad Med 70: 203–208

Doerr HW, Gross G, Schmitz H (1976) Neutralising serum IgM antibodies in infection with herpes simplex virus hominis. Med Microbiol Immunol (Berl) 162: 183–192

Douglas JM, Critchlow C, Benedetti J et al. (1984) A double blind study of oral acyclovir for suppression of recurrences of genital herpes simplex virus infection. N Engl J Med 310: 1551–1556

Drew WL, Mintz L., Miner RC, Sands M, Ketterer B (1981) Prevalence of cytomegalovirus infection in homosexual men. J Infect Dis 143: 188–192

Elion GB (1982) Mechanism of action and selectivity of acyclovir. Am J Med 73 [Suppl 1A]: 7–13

Embil JA, Garner JB, Pereira LH, White FMM, Manuel FR (1985) Association of cytomegalovirus and herpes simplex virus infections of the cervix in four clinic populations. Sex Trans Dis 12: 224–227

Evans AS (1978) Infectious mononucleosis and related syndromes. Am J Med Sci 276: 325–339

Fiddian AP, Kinghorn GR, Goldmeier D et al. (1983) Topical acyclovir in the treatment of genital herpes: a comparison with systemic therapy. J Antimicrob Chemother 12 [Suppl 3]: 67–77

Figuero ME, Rawls WE (1969) Biological markers for differentation of herpes virus strains of oral and genital origin. J Gen Virol 4: 259–267

Goldmeier D. (1980) Proctitis and herpes simplex virus in homosexual men. Br J Vener Dis 56: 111–114

Goldmeier D, Bateman JRM, Rodin P (1975) Urinary retention and intestinal obstruction associated with ano-rectal herpes simplex virus infection. Br Med J 1: 425–426

Goodell SE, Quinn TC, Mkrtichian E, Schuffler MD, Holmes KK, Corey L (1983) Herpes simplex virus proctitis in homosexual men. Clinical, sigmoidoscopic and histopathological features. N Engl J Med 308: 868–871

Griffiths PD, Panjwani DD, Stirk PR et al. (1984) Rapid diagnosis of cytomegalovirus infection in immunocompromised patients by detection of early antigen fluorescent foci. Lancet ii: 1242–1244

Gudgel EF (1954) Can molluscum contagiosum be a venereal disease? US Armed Forces Med J 5: 1207–1208

Guinan ME, MacCalman J, Kern ER, Overall JC, Spruance SL (1981) The course of untreated recurrent genital herpes simplex infection in 27 women. N Engl J Med 304: 759–763

Hill TJ (1985) Herpes simplex virus latency. In: Roizman B (ed) The herpes viruses, vol. 3. Plenum Press, New York, pp 175–240

Hilleman MR, Larson VM, Lehman ED et al. (1981). Subunit herpes simplex virus 2 vaccine. In: Nahmias AJ, Dowdle WR, Schinazi RF (eds) The human herpesviruses: an interdisciplinary approach. Elsevier, New York, pp 503–506

Hindley DJ, Adler MW (1985) Genital herpes: an increasing problem? Genitourin Med 61: 56–58

Jacobs PH (1970) Molluscum contagiosum. Aerosp Med 41: 1196–1197

Jacome DE, Yanez GF (1980) Herpes genitalis and neurogenic bladder and bowel. J Urol 124: 752

Jeansson S, Molin L (1974) On the occurrence of genital herpes simplex virus infection. Acta Dermatol 54: 479–485

Jordan CM, Rousseau WE, Noble GR, Stewart JA, Chin TDY (1973) Association of cervical cytomegaloviruses with venereal disease. N Engl J Med 288: 932–934

Kelman AD, Capozza FE, Kibrick S (1975) Differential action of deoxynucleosides on mammalion cell culture infected with herpes simplex virus type 1 and 2. J Infect Dis 131: 452–455

Klastersky J (1977) Ascending myelitis in association with herpes simplex virus. N Engl J Med 297: 182–184

Klemola E, Strenstrom R, Von Essen R (1972) Pneumonia as a clinical manifestation of cytomegalovirus infection in previously healthy adults. Scan J Infect Dis 4: 7–10

Krech U (1973) Complement fixing antibodies against cytomegalovirus in different parts of the world. Bull WHO 49: 103

Kristensson K (1978) Retrograde transport of macromolecules in axons. Annu Rev Pharmacol Toxicol 18: 97–110

Lamb SG, Stern H (1966) Cytomegalovirus mononucleosis with jaundice as presenting sign. Lancet ii: 1003–1006

Lang DJ, Kummer JF (1972) Demonstration of cytomegalovirus in semen. N Engl J Med 287: 756–758

Lasky LA, Dowbenko, Simonsen CC, Berman PW (1984) Protection of mice from lethal herpes simplex virus infection by vaccination with a secreted form of cloned glycoprotein. Biotechnol 2: 527–532

Lee FK, Coleman RM, Pereira L, Bailey PD, Tatsuno M, Nahmias AJ (1985) Detection of herpes simplex virus type 2, specific antibody with glycoprotein G. J Clin Microbiol 22: 641–644

Lee FK, Pereira, Griffin C, Reid E, Nahmias A (1986) A novel glycoprotein for detection of herpes simplex virus type 1 specific antibodies. J Virol Methods 14: 111–118

Leiden JM, Buttyan R, Spear PG (1976) Herpes simplex virus gene expression in transformed cells. 1. Regulation of the viral thymidine kinase gene in transformed L cells by products of superinfecting virus. J Virol 20: 413–424

Lonsdale DM (1979) A rapid technique for distinguishing herpes simplex virus type 1 from type 2 by restriction enzyme technology. Lancet I: 849–852

Lory D, Madara TJ, Ponce de leon M, Cohen GH, Montgomery PC, Eisenberg RJ (1984) Glycoprotein D protects mice against lethal challenge with herpes simplex virus types 1 and 2. Infect Immunol 42: 761–764

Luby ED, Gillespie O (1981) Psychological responses to genital herpes. The Helper 3(4): 2–3

Lycke E, Kristensson K, Svennaholm B, Vahlne A, Ziegler R (1984) Uptake and transport of herpes simplex virus in neurites of rat dorsal root dorsal ganglia cells in culture. J Gen Virol 65: 55–64

Manne S, Sandler I (1984) Coping and adjustment to genital herpes. J Behav Med 7: 391–410

McSwiggan DA, Darougar S, Rahman AF, Gibson JA (1975) Comparison of the sensitivity of human embryo kidney cells, Hela cells and W1-38 cells for the primary isolation of viruses from the eye. J Clin Pathol 28: 410–413

Mertz GJ, Critchlow CW, Benedetti J et al. (1984a) Double blind placebo controlled trial of oral acyclovir in first episode genital herpes simplex virus infection. JAMA 252: 1147–1151

Mertz GJ, Peterman G, Ashley R et al. (1984b) Herpes simplex virus type-2 glycoprotein subunit vaccine: tolerance and humoral and cellular responses in humans. J Infect Dis 150: 242–249

Mertz GJ, Schmidt O, Jourden JL et al. (1985) Frequency of acquisition of first episode genital infection with herpes simplex virus from symptomatic and asymptomatic source contacts. Sex Trans Dis 12: 33–39

Meyer HM, Johnson RT, Crawford IP, Dascomb HE, Rogers NG (1960) Central nervous system syndromes of "viral" etiology: a study of 713 cases. Am J Med 29: 334–347

Mindel A, Adler MW (1984) Genital herpes. Gower Medical, London

Mindel A, Sutherland S (1984) Antibodies to cytomegalovirus in homosexual and heterosexual men attending an STD clinic. Br J Vener Dis 60: 189–192

Mindel A, Adler MW, Sutherland S, Fiddian AP (1982) Intravenous acyclovir treatment for primary genital herpes. Lancet i: 697–700

Mindel A, Weller IVD, Faherty A et al. (1984) Prophylactic oral acyclovir in recurrent genital herpes. Lancet ii: 57–59

Montgomerie JZ, Becrof DMO, Croxon MC, Doak PB, North JDK (1969) Herpes simplex virus infection after renal transplantation. Lancet 2: 867–870

Muller SA, Herrmann EC, Winkelmann RK (1972) Herpes simplex infection in hematologic malignancies. Am J Med 52: 102–114

Nahmias AJ, Dowdle WR (1968) Antigenic and biologic differences in herpes virus hominis. Prog Med Virol 10: 110–159

Nahmias AJ, Roizman B (1973a) Infection with herpes simplex viruses 1 and 2 (First of three parts). N Engl J Med 289: 667–674

Nahmias AJ, Roizman B (1973b) Infection with herpes simplex viruses 1 and 2 (Third of three parts). N Engl J Med 289: 751–759

Nahmias AJ, Josey WE, Naib ZM, Luce CF, Duffey A (1970) Antibodies to herpes virus hominis type 1 and 2 in humans. Am J Epidemiol 91: 539–546

Nahmias AJ, delBuono I, Pipkin J, et al. (1971) Rapid identification and typing of herpes simplex virus type I and II by a direct immunofluorescent technique. Appl microbiol 22: 455–458

Naib ZM (1966) Exfoliative cytology of viral cervico-vaginitis. Acta Cytol (Baltimore) 10: 126–129

Nilsen AE, Aasen T, Halsos AM et al. (1982) Efficacy of oral acyclovir in the treatment of initial and recurrent genital herpes. Lancet ii: 571–573

Oates JK, Greenhouse PRDH (1978) Retention of urine in anogenital herpetic infection. Lancet i: 691–692

Powell KL, Purifoy DJ (1977) Nonstructural proteins of herpes simplex virus I: purification of the induced DNA polymerase. J Virol 24: 618–626

Powers RD, Rein MF, Hayden FG (1982) Necrotizing balanitis due to herpes simplex type 1. JAMA 248: 215–216

Puga A, Rosenthal, JD, Openshaw H, Notkins AL (1978) Herpes simplex virus DNA and in RNA sequences in acutely and chronically infected trigeminal ganglia of mice. Virology 89: 102–111

Rapp F (1984) Herpes simplex viruses. In: Holmes KK, Marh PA, Sparling DF, Weisner P (eds) Sexually transmitted diseases. McGraw Hill, New York, pp 438–449

Rawls WE (1979) Herpes simplex virus types 1 and 2 and herpesvirus simiae. In: Lennette EH, Schmidt NJ (eds) Diagnostic procedures for viral, rickettsial and chlamydial infection. American Public Health Association, Washington DC, pp 309–373

Rawls WE, Campione-Piccardo J (1981) Epidemiology of herpes simplex viruses type 1 and 2 infections. In: Nahmias AJ, Dowdle WR, Schinazi F (eds) The human herpesviruses: an interdisciplinary approach. Elsevier, New York, pp 137–152

Reeves WC, Corey L, Adams HG, Vontver LA, Holmes KK (1981) Risk of recurrence after first episode of genital herpes. Relation to HSV type and antibody response. N Engl J Med 305: 315–319

Reichman RC, Badger GJ, Guinan ME et al. (1983) Topically administered acyclovir in the treatment of recurrent herpes simplex genitalis. A controlled trial. J Infect Dis 147: 336–340

Reichman RC, Badger GJ, Mertz GJ et al. (1984) Treatment of recurrent genital herpes simplex infection with oral acyclovir. A controlled trial. JAMA 251: 2103–2107

Riehle RA, Williams JJ (1979) Transient neuropathic bladder following herpes simplex genitalis. J Urol 122: 263–264

Roizman B, Frenkel N (1973) The transcription and state of herpes simplex virus DNA in productive infection and human cervical cancer tissue. Cancer Res 33: 1402–1416

Roizman B, Furlong D (1974) The replication of herpesviruses. In: Frenkel-Conrat H, Wagner RR (eds) Comprehensive virology, vol. 3. Plenum Press, New York, pp 63–141

Salo OP, Lassus A, Hovi T, Fiddian AP (1983) Double blind placebo controlled trial of oral acyclovir in recurrent genital herpes. Eur J Sex Transm Dis 1: 95–98

Samarasinghe PL, Oates JK, MacLennan IPB (1979) Herpetic proctitis and sacral radiomyelopathy – a hazard for homosexual men. Br Med J 286: 365–366

Schmitz H, Enders G (1977) Cytomegalovirus as a frequent cause of Guillain–Barré syndrome. J Med Virol 1: 21–27

Siegal FP, Lopez C, Hammer GS et al. (1981) Severe acquired immunodeficiency in male homosexuals, manifested by chronic perianal ulcerative herpes simplex lesions. N Engl J Med 305: 1439–1444

Simmons A, Nash AA (1984) Zosteriform species of herpes simplex virus as a model of recrudescence and its use to investigate the role of immune cells in prevention of recurrent disease. J Virol 52: 816–821

Skinner GRB, Woodman CBJ, Hartley CE et al. (1982) Preparation and immunogenicity of vaccine AcNFU,(S⁻) MRC towards the prevention of herpes genitalis. Br J Vener Dis 58: 381–6

Skoldenberg B, Jeansson S, Wolonitis S (1975) Herpes simplex virus type 2 and acute aseptic meningitis: clinical features of cases with isolation of herpes simplex virus from cerebrospinal fluids. Scan J Infect Dis 7: 227–232

Spear PG (1984) Glycoprotein specified by herpes simplex virus. In: Roizman B (ed) The herpesviruses, vol 3. Plenum Press, New York, pp 315–356

Spruance SL, Overall JC, Kern ER, Krueger GC, Pliam V, Miller W (1977) The natural history of recurrent herpes simplex labialis. N Engl J Med 297 69–75

Stern H, Elek SD (1965) The evidence of infection with cytomegalovirus in a normal population. A serological study in Greater London. J London 63: 79–87

Stevens JG (1978) Latent characteristics of selected herpesviruses. Adv Cancer Res 26: 227–256

Straus SE, Takiff HE, Seidlin M et al. (1984) Suppression of frequently recurring genital herpes. A placebo-controlled double blind trial of oral acyclovir. N Engl J Med 310: 1545–1550

Thin RN, Nabaro JM, Parker JD, Fiddian AP (1983) Topical acyclovir in the treatment of initial genital herpes. Br J Vener Dis 59: 116–119

Tottie M (1943) Herpes genitalis subsequente bubonulo. Acta Dermatol Venereol 23: 306–309

Vahlne A, Nystrom B, Sandberg M, Hamberger A, Lycke E (1978) Attachment of herpes simplex virus to neurons and glial cells. J Gen Virol 40: 359–371

Vestergaard BF, Graballe PC (1979) ELISA for herpes simplex virus (HSV) type specific antibodies in human sera using HSV type 1 and type 2 polyspecific antigens blocked with type-heterologous rabbit antibodies. Acta Pathol Microbiol Scand Ser B 87: 261–263

Wagner EK (1985) Individual HSV transcripts: characterisation of specific genes. In: Roizman B (ed) The herpesviruses, vol, 3. Plenum Press, New York, pp 45–104

Waugh MA (1976) Anorectal herpesvirus hominis infection in men. J Am Vener Dis Assoc 31: 68–70

Wentworth BB, Alexander ER (1971) Seroepidemiology of infections due to members of the herpesvirus group. Am J Epidemiol 94(5): 496–507

Wildy P (1985) Herpes viruses: a background. In: Tyrrell DAJ, Oxford JS (eds) Antiviral chemotherapy and interferon. Br Med Bull 41: 3339–3441

Wildy P, Russell WC, Horne RW (1960) The morphology of herpes virus. Virology 12: 1044–1052

Wildy P, Field HJ, Nash AA (1982) Classical herpes latency revisited. In: Mahy BWJ, Minson AC, Darby GK (eds) Virus persistence. Cambridge University Press, Cambridge, pp 133–167

Wilkin JK (1977) Molluscum contagiosum venereum in a women's outpatient clinic: a venereally transmitted disease. Am J Obstet Gynecol 128: 531–535

Wilson RSE, Morris TH, Rees JR (1972) Cytomegalovirus myocarditis. Br Heart J 34: 865–868

Chapter 4

Protozoal Infections

E. Allason-Jones

Introduction

In recent years, attention has been drawn to the association between male homosexuality and a variety of enteric infections. Multiple sexual partners, specific sexual practices and a large reservoir of asymptomatic carriers all contribute to the spread of infection among homosexual men.

Patients with gastrointestinal symptoms may not associate their problems with sexual activities, or be unwilling to admit to homosexuality. Consequently they present to a variety of specialists for investigation. Medical staff need to be aware of the possibility of sexually transmitted infection and to be prepared to ask the patient about his sexual orientation. Failure to do so may lead to the significance of the results being misinterpreted. To complicate matters further, the clinical significance of some infections may be radically altered if the patient's immune status is compromised.

The protozoal organisms that are of importance in the context of homosexuality are *Entamoeba histolytica* and *Giardia lamblia*, and in patients with the acquired immune deficiency syndrome (AIDS) *Cryptosporidium*, *Isospora belli* and more recently the *Microsporidia*.

Entamoeba histolytica

Amoebiasis is generally considered to be a disease of tropical areas where public health measures are insufficient to prevent transmission of *Entamoeba histolytica*, the causative organism. Amoebic dysentery and other complications of invasive disease are a major cause of morbidity and mortality in these areas.

Until recently, in temperate zones such as the United Kingdom, *E. histolytica* was found almost exclusively in travellers returning from endemic areas. However, the organism may commonly be found in the stools of homosexual men who have never travelled to high-risk areas. Contrary to the situation in the tropics, invasive disease is very rare amongst these patients, the vast majority being asymptomatic cyst-passers. As a result controversy has developed over the need to treat these individuals.

Prevalence of *E. histolytica* in Homosexual Men

The possibility that *E. histolytica* could be sexually transmitted was first raised in 1968, when the parasite was reported in three homosexual men, living in New York, who were sexual partners of a man recently returned from India (Most 1968). A number of reports later appeared in which *E. histolytica* found in the stools of homosexual men was thought to be responsible for the patients' symptoms (Kazal et al. 1976; Kean 1976; Mildvan et al. 1977; Schmerin et al. 1977; Hurwitz and Owen 1978). However, only three reports have provided conclusive evidence of invasive disease (Burnham et al. 1980; Ylvisaker and McDonald 1980; Saltzberg and Hall-Craggs 1986).

In contrast to the rare cases of invasive amoebiasis, cyst excretion is a common finding in homosexuals (Table 4.1). Prevalence studies in North America have shown that 20%–32% of selected homosexuals screened in sexually transmitted disease (STD) clinics were passing *E. histolytica* cysts (William et al. 1978; Kean et al. 1979; Keystone et al. 1980; Phillips et al. 1981; Markell et al. 1984; Sorvillo et al. 1986). Of heterosexual patients screened, 0–1% were similarly infected. Reports on the prevalence of *E. histolytica* among homosexuals in Germany, Sweden and Finland have shown 21%–23% of those screened to be infected (Bienzle et al. 1984; Hakansson et al. 1984; L. Jokipii et al. 1985b). In the UK prevalence studies have shown that 11%–20% of homosexual men attending STD clinics are carrying *E. histolytica* cysts; no cases have been found among heterosexual control groups (Sargeaunt et al. 1983; Chin and Gerken 1984; Allason-Jones et al. 1986).

Whilst many of these studies are open to criticism over details of patient selection, comparative groups or specimen handling, there seems little doubt that *E. histolytica* is endemic in many homosexual populations in areas where the organism would usually be considered a rarity. The true prevalence and potential numbers of individuals infected are difficult to establish with any certainty, since homosexuals attending genito-urinary medicine clinics are not necessarily representative of the homosexual population at risk.

Transmission of *E. histolytica* Among Homosexual Men

A number of studies provide evidence to support the suggestion that oral–anal sex is the mode of spread of infection. Quinn et al. (1983) found a positive correlation between the practice of anilingus and the presence of *E. histolytica* cysts, and noted that 70% of their survey population of 194 homosexuals had engaged in oral–anal sex in the month prior to enrolling in the survey. Phillips et al. (1981) also found a significant association between *E. histolytica* infection and oral–anal sex, which was independent of the number of sexual partners. The only UK study to enquire about sexual activities reports the same correlation between protozoal cyst carriage and

Table 4.1. Published prevalences of *Entamoeba histolytica* in homosexual men and comparative groups

	Homosexual men No. infected/ no. screened	(%)	Comparative groups No. infected/ no. screened	(%)	Reference
North America					
New York	$18_{/89}$	(20)	$4_{/139}$	(3)	William et al. (1978)
New York	$39_{/126}$	(31)	$74_{/5885}$	(1)	Kean et al. (1979)
Toronto	$54_{/200}$	(27)	$1_{/100}$	(1)	Keystone et al. (1980)
New York	$10_{/51}$	(20)	$0_{/64}$	(0)	Phillips et al. (1981)
San Francisco	$145_{/508}$	(29)	—	—	Markell et al. (1984)
Los Angeles	$38_{/140}$	(27)	—	—	Sorvillo et al. (1986)
Europe					
UK	$52_{/470}$	(11)	—	—	Sargeaunt et al. (1983)
Berlin	$45_{/197}$	(23)	—	—	Bienzle et al. (1984)
London	$10_{/83}$	(12)	$0_{/43}$	(0)	Chin and Gerken (1984)
Goteborg	$28_{/133}$	(21)	$0_{/27}$	(0)	Hakansson et al. (1984)
Helsinki	$36_{/153}$	(24)	$2_{/119}$	(2)	L. Jokipii et al. (1985b)
London	$45_{/225}$	(20)	$0_{/129}$	(0)	Allason-Jones et al. (1986)

Not all comparative groups have been drawn from STD clinics and may not be exclusively heterosexual.

oral–anal sex and found that 50% of homosexuals admitted to the practice (Chin and Gerken 1984). No association has been reported between the presence of intestinal parasites and other sexual activities such as receptive or insertive anal intercourse. Infection has not been shown to correlate with foreign travel and would not explain the significant difference between homosexual and heterosexual prevalence rates.

Clinical Significance of *E. histolytica*

Invasive Amoebiasis

Infection usually results from ingesting food or water contaminated with *E. histolytica* cysts. The cysts are not destroyed in the stomach but pass through to the small intestine where they release trophozoites. Pathogenic trophozoites invade the intestinal mucosa causing mucosal ulceration. The symptoms that result can be mild, with abdominal discomfort, flatulence and loose stools, or severe with marked abdominal pain and watery diarrhoea that contains blood and mucus. Intestinal perforation with subsequent peritonitis and septicaemia may develop. Amoebae can sometimes reach the liver via the portal vein and result in amoebic liver "abscess". Much less commonly, cutaneous and genital lesions can develop and cerebral and pulmonary lesions have been described.

Few cases of invasive amoebiasis have been described in homosexual men. One case of amoebic dysentery has been reported in a homosexual man in the UK, his sexual partner was also infected (Burnham et al. 1980). Two separate cases of amoebic colitis with liver abscess have been reported in North America (Ylvisaker and McDonald 1980). Recently, a fatal case of amoebic colitis in a schizophrenic man was described, the patients psychiatric condition preventing a diagnosis being reached until post mortem (Saltzberg & Hall-Craggs 1986).

Diagnosis of Invasive Amoebiasis

The diagnosis of invasive amoebiasis is based on three criteria: haematophagous trophozoites in the faeces, typical endoscopic or rectal biopsy findings, and high titre serum antibodies. The demonstration of any one of these three is sufficient to make the diagnosis.

Conclusive evidence of invasive disease is provided by finding haematophagous trophozoites of E. histolytica in the stools. No other amoebic species ingest red blood cells, nor do E. histolytica trophozoites that are not invading the host tissues. Trophozoites are often scanty and die rapidly once outside the human body, hence the need for immediate examination of faecal specimens by an experienced parasitologist.

Sigmoidoscopy classically shows multiple small ulcers that have swollen mucosal edges and are covered in a white or yellow exudate. The intervening mucosa often appears to be normal, although a more diffuse granular inflammation has been described. Histologically, marked inflammatory oedema with scant leucocytic infiltration is seen in the submucosa. Ulceration of the overlying mucosa develops as tissue necrosis progresses. Entamoeba histolytica trophozoites are frequently seen in histological examination of the ulcer bases and mucosal exudate.

Serological tests for E. histolytica antibodies are positive in over 75% of patients with invasive amoebiasis and in an even greater percentage of patients with extra-intestinal disease. A number of techniques are available for the detection of E. histolytica antibodies, but since positive titres may persist for some time after treatment they cannot differentiate between active and past infections. Antigen-detection techniques that would overcome this problem are being assessed.

Cyst Carriage

The majority of individuals infected with E. histolytica do not develop dysentery or any other of the complications described. With a normal intestinal transit time, faeces become increasingly solid as they pass along the colon and so the trophozoites encyst to survive. As a result cysts, but rarely trophozoites, are found in stool samples.

Cyst carriage has been blamed for a variety of minor gastrointestinal symptoms such as vague abdominal pain, diarrhoea and flatulence. However, several studies among homosexual men have consistently reported a lack of any correlation between symptoms and E. histolytica cyst carriage (William et al. 1978; Keystone et al. 1980; Bienzle et al. 1984; Chin and Gerken 1984; Markell et al. 1984; L. Jokipii et al. 1985b; Allason-Jones et al. 1986). Quinn et al. (1983) compared a group of 119 homosexual men complaining of ano-rectal or intestinal symptoms with 75 asymptomatic homosexuals and found that a similar percentage of the two groups was infected with E. histolytica. Several studies have noted that homosexual men are more likely than heterosexual men to complain of a variety of gastrointestinal symptoms, although no correlation could be found between symptoms and the presence of E. histolytica cysts

(Keystone et al. 1980; Chin and Gerken 1984; Allason-Jones et al. 1986). Only one study found an increased likelihood of blood or mucus in the stools of patients with *E. histolytica* (but no increase in complaints of diarrhoea, abdominal cramp or flatulence); however, they had not excluded the possibility of other pathogens (Phillips et al. 1981). The reason for the lack of correlation between cyst carriage and symptoms may relate to the recent observation that some *E. histolytica* strains (or zymodemes) are non-pathogenic.

Zymodeme Classification

Entamoeba histolytica isolates have been classified into zymodemes on the basis of their isoenzymes (Sargeaunt and Williams 1978). The technique involves culturing cysts or trophozoites from the patients' faeces and subjecting them to thin-layer starch gel electrophoresis (Fig. 4.1). Twenty-two different zymodemes have now been demonstrated, 12 of which have been found only in asymptomatic cyst-passers and are therefore considered to be non-pathogenic (Sargeaunt et al. 1987). Nine other zymodemes have been isolated from patients with invasive disease and so are of proven pathogenicity. The position of one zymodeme remains uncertain (Sargeaunt et al. 1978, 1980, 1982a, b, c, d, 1984; Jackson et al. 1982; Gathiram and Jackson 1985).

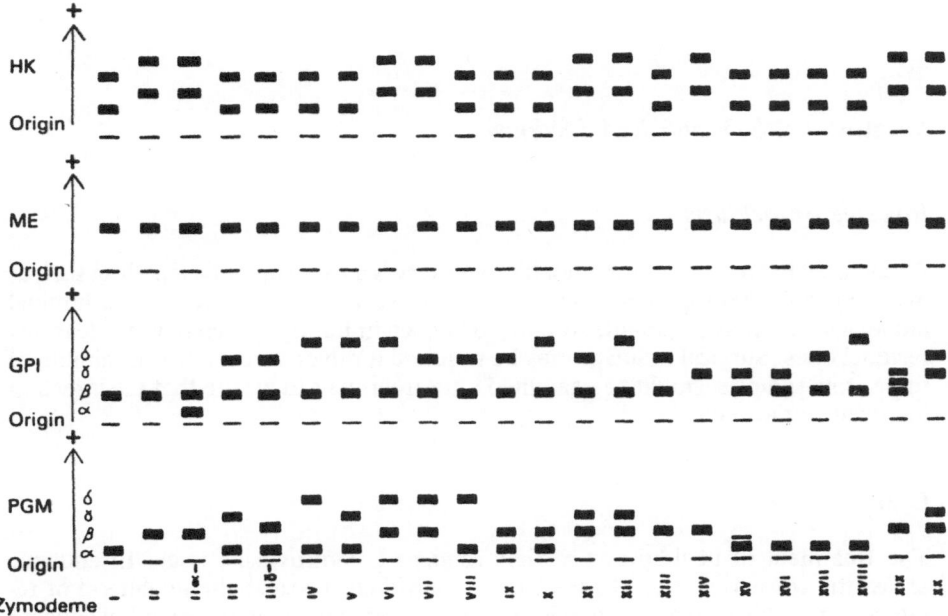

Fig. 4.1. Zymodemes of *Entamoeba histolytica* identified using EC 5319 glucose phosphate isomerase (*GPI*); EC 11140 L-malate:NADP+ oxidoreductase (oxaloacetate decarboxylating) (*ME*); EC 2751 phosphoglucomutase (*PGM*); and EC 2711 hexoknase (*HK*).

A zymodeme is a population of amoebae differing from similar populations in the electrophoretic mobility of certain enzymes. The markers for pathogenicity are the absence of the α-band together with the presence of the β-band in *PGM*. Advanced bands in *HK* confirm the *PGM* results. The only exception is zymodeme XIII, which lacks advanced *HK* bands (from Sargeaunt 1985).

Patients infected with a pathogenic zymodeme, even if asymptomatic at the time, are often strongly seropositive for *E. histolytica* antibodies, suggesting that occult tissue damage may be occurring. Conversely, patients with non-pathogenic zymodemes have been shown to have negative or only weakly positive serology, supporting the suggestion that in these cases the parasite is behaving as a commensal (Jackson et al. 1985).

Zymodeme classification has been carried out on *E. histolytica* isolates from homosexual men in three UK studies (Sargeaunt et al. 1983; Allason-Jones et al. 1986; Goldmeier et al. 1986). Of the total 109 classified isolates, all have been non-pathogenic zymodemes. Fifty-eight of these patients have been further investigated; all have had negative amoebic serology and no clinical evidence of invasive amoebiasis (McMillan et al. 1984; Goldmeier et al. 1986). Little information is available on the zymodemes present in homosexual men in other countries, but of 18 such isolates studied in North America all were classified as non-pathogenic (Mathews et al. 1986).

Zymodeme classification offers the potential for a rational treatment policy to be devised for *E. histolytica* cyst-carriers. Because the technique is time consuming and expensive it is unlikely to become available on a routine basis; classification of a single isolate costs about £25–£50. Optimal use of zymodeme classification will be in epidemiological studies identifying populations where pathogenic zymodemes are common and treatment likely to be of benefit and in those where non-pathogenic zymodemes predominate and treatment is unnecessary.

Treatment of *E. histolytica* infections

Invasive Amoebiasis

Recommended treatment for amoebic dysentery is a tissue amoebicidal drug such as metronidazole 800 mg eight hourly for five to seven days, followed by a luminal amoebicide such as diloxanide furoate 500 mg eight hourly for ten days to clear any residual cysts. Surgical drainage may be required for liver abscesses. A minimum of three stool samples should be examined after treatment to ensure that the infection has been eradicated.

Cyst Carriage

The treatment of healthy cyst-carriers remains a controversial issue. In endemic areas, the consensus is not to treat these individuals because the likelihood of re-infection is so high. In non-endemic areas they tend to be treated because of concern about the possible development of invasive disease or transmission to others. In the absence of any of the accepted features of invasive amoebiasis, there is no evidence to suggest that treatment of homosexual men found to be passing *E. histolytica* cysts is of any benefit. Cyst carriage is not associated with gastrointestinal symptoms, and invasive disease has never been documented in carriers of the non-pathogenic zymodemes that are circulating in the homosexual community.

E. *histolytica* and the Acquired Immune Deficiency Syndrome

Further research is needed on the prevalence and pathogenicity of *E. histolytica* in patients with AIDS. It remains to be seen whether non-pathogenic zymodemes are capable of causing clinical disease in immunocompromised hosts; as yet there is no evidence that they are. Given the number of homosexual men who have developed AIDS, there are remarkably few reports of *E. histolytica* in these patients and none has provided evidence of invasive disease; in symptomatic patients, other pathogens have coexisted. Suggestions that infection with non-pathogenic *E. histolytica* could cause stimulation of T cells infected with human immunodeficiency virus (HIV) and thereby accelerate the development of AIDS in these patients are purely speculative (Pearce 1983).

Other Amoebic Species

There are a number of other amoebic species that can infect the intestinal tract. These are of academic interest only as they are not associated with disease. They are found commonly in the faeces of homosexual men and are presumably spread by faecal–oral contamination, such as occurs during oral–anal sex. *Entamoeba coli*, *Entamoeba hartmanni* and *Endolimax nana* are the most commonly found; prevalence rates amongst homosexual men in one UK study were 20%, 14% and 24%, respectively (Allason-Jones et al. 1986). Prevalence rates for heterosexual men in the same study were less than 5% in each case. Similar results have been reported in American studies. *Iodamoeba buetschlii* and *Dientamoeba fragilis* are less commonly found. The latter protozoan does not encyst and as trophozoites die rapidly outside the bowel they will only be found in fresh stool specimens. There has been some debate over the possibility that *D. fragilis* can cause a mild, self-limiting diarrhoea but tissue damage has never been proven. Treatment is not indicated for any of these organisms.

Giardia lamblia

Giardia lamblia is a flagellate protozoan that infects the upper small intestine of man. Infection results from ingesting cysts, usually in contaminated food or water. Giardiasis occurs worldwide and is a common cause of travellers diarrhoea. Sporadic outbreaks of infection have occurred in the UK.

Prevalence of *G. lamblia* in Homosexual Men

In the 1970s a handful of cases of giardiasis in homosexual men were reported in the USA (Meyers et al. 1977; Mildvan et al. 1977; Hurwitz and Owen 1978; Schmerin et al. 1978). Prevalence studies in North America have subsequently shown 4%–18% of homosexual men screened to be infected with *G. lamblia* (Table 4.2) (William et al. 1978; Kean et al. 1979; Keystone et al. 1980; Phillips et al. 1981; Markell et al. 1984; Sorvillo et al. 1986). However, the populations studied have been biased, usually volunteers attending STD clinics. Comparative groups have had prevalence rates of 0–3%. Only one stool sample from each patient was examined, and, given the erratic pattern of *G. lamblia* cyst excretion, this almost certainly means the results are an underestimate of the number of patients infected. Prevalence studies in Germany, Sweden and Finland have shown 6%–9% of homosexuals screened to be infected with *G. lamblia* (Bienzle et al. 1984; Hakansson et al. 1984; L. Jokipii et al. 1985b).

Table 4.2. Published prevalences of *Giardia lamblia* in homosexual men and comparative groups

	Homosexual men No. infected/ no. screened	(%)	Comparative groups No. infected/ no. screened	(%)	Reference
North America					
New York	$11/89$	(12)	$4/139$	(3)	William et al. (1978)
New York	$23/126$	(18)	$123/5885$	(2)	Kean et al. (1979)
Toronto	$26/200$	(13)	$3/100$	(3)	Keystone et al. (1980)
New York	$2/51$	(4)	$0/64$	(0)	Phillips et al. (1981)
San Francisco	$29/508$	(6)	—	—	Markell et al. (1984)
Los Angeles	$22/140$	(16)	—	—	Sorvillo et al. (1986)
Europe					
UK	$17/470$	(4)	—	—	Sargeaunt et al. (1983)
Berlin	$18/197$	(9)	—	—	Bienzle et al. (1984)
London	$7/83$	(8)	$0/43$	(0)	Chin and Gerken (1984)
Goteborg	$10/133$	(8)	$0/27$	(0)	Hakansson et al. (1984)
Helsinki	$10/153$	(7)	$15/119$	(13)	L. Jokipii et al. (1985b)
London	$7/225$	(3)	$3/129$	(2)	Allason-Jones et al. (1986)

Not all comparative groups have been drawn from STD clinics and may not be exclusively heterosexual.

Prevalence studies in the UK also have problems with patient selection and single-stool examinations, and the small numbers identified cause difficulties with statistical interpretation (Sargeaunt et al. 1983; Chin et al. 1984; Allason-Jones et al. 1986). Of homosexual men screened, 3%–8% were found to be infected with *G. lamblia*, whereas 0–2% of heterosexual controls were similarly infected. Not all studies have shown a statistically significant difference in prevalence between the homosexual and heterosexual groups.

Transmission of *G. lamblia* among homosexual men appears to be associated with oral–anal sexual contact, reflecting the faecal–oral contamination that occurs (Quinn et al. 1983; Markell et al. 1984). Only a minute amount of faeces would need to be ingested as the infecting dose of *G. lamblia* is small; prisoner volunteers fed cysts of *G. lamblia* all became infected after ingesting 100 cysts and, in some, 10 cysts were sufficient to establish infection (Rendtorff 1954).

Clinical Significance of *G. lamblia*

There is a considerable range in the severity of symptoms associated with *G. lamblia* infection. Classically, symptoms develop two weeks after infection, commencing with diarrhoea of sudden onset. The stools are loose or watery, yellowish in colour and have an offensive smell. Abdominal distension, discomfort and flatulence are common. Some patients have a more insidious change in bowel habit or more protracted symptoms. In these cases, tiredness, weight loss and evidence of malabsorption may predominate. The majority of patients infected with *G. lamblia* remain completely asymptomatic throughout.

Homosexual men infected with *G. lamblia* can present with any of the features described above. Asymptomatic infection appears to be common; several studies have reported a lack of correlation between infection and gastrointestinal symptoms (William et al. 1978; Keystone et al. 1980; Phillips et al. 1981; Bienzle et al. 1984; Chin and Gerken 1984; Allason-Jones et al. 1986).

Pathogenesis

The mechanism by which *G. lamblia* causes disease is not clear. The organism excysts in the duodenum to release trophozoites that attach to the intestinal epithelium by means of a ventral sucking disc. Mucosal invasion by trophozoites is rare, but electron microscopy has shown the epithelial brush borders beneath the trophozoites to be damaged. Immunocytes accumulate in the lamina propria and possibly the immunological response of the host contributes to mucosal damage. Abnormal bacterial flora have been found in patients with malabsorption but it seems unlikely that these play a primary role in pathogenesis.

Diagnosis

The diagnosis is made by finding cysts on microscopy. Trophozoites are unlikely to be found unless diarrhoea is severe. A number of samples may need to be examined to make the diagnosis, as cyst excretion varies from day to day. Techniques for the detection of *Giardia* antigen in the faeces are being developed, which may have useful diagnostic applications (Green et al. 1985). Sampling of jejunal fluid or examination of jejunal biopsy material may be necessary if giardiasis is suspected and cysts cannot be found in the stools. Serological tests for IgG antibodies to *G. lamblia*, which are not generally available, may be positive in up to 80% of symptomatic patients.

Treatment

Metronidazole and tinidazole are both highly effective in the treatment of giardiasis. Other drugs such as mepacrine and chloroquine have no particular advantage and tend to cause more side-effects. Metronidazole given as a single dose of 2 g daily for three successive days will clear the parasite in the majority of cases. Patients unable to tolerate high doses may be given 200 mg three times daily for 14 days. Most symptomatic patients will have a rapid response to treatment, although some with an acquired lactose intolerance will have a more gradual resolution of symptoms.

The treatment of asymptomatic carriers remains a cause of controversy. Most cases are treated, once detected, to prevent further transmission of the parasite. Spontaneous eradication of *G. lamblia* is probably a very common occurrence (Rendtorff 1954).

Cryptosporidium

The coccidian protozoan parasite *Cryptosporidium* was first described in mice in 1907 (Tyzzer 1907), and has since been found in a variety of domestic and farm animals (Pohlenz et al. 1978; Angus 1983). The first report of *Cryptosporidium* in man appeared in 1976, when the organism was found in a child, living on a farm, who developed diarrhoea (Nime et al. 1976). In view of the veterinarian interest, and because many of the initial reports involved farm workers or other animal handlers, the infection was thought to be a zoonosis. It now seems that person-to-person transmission, or contaminated food or water supplies play a major role in the spread of infection. Additional interest in *Cryptosporidium* as a human pathogen developed when it became apparent that the organism was capable of causing profuse life-threatening diarrhoea in immunocompromised patients.

The Organism

The life-cycle of *Cryptosporidium* is complex and not fully understood. Oocysts are ingested and release sporozoites, which take up residence on enterocytes on the microvillous border of the intestinal epithelium. Sporozoites mature into trophozoites, the active feeding form, and these undergo asexual budding to produce schizonts, which release merozoites. These merozoites can in turn infect epithelial cells, mature into trophozoites and give rise to further merozoites. Second-generation merozoites produce the microgametes and macrogametes of the sexual phase, which fuse to form the zygote. The zygote secretes a wall around itself thus transforming into a oocyst. All of these developmental forms can be found in faecal specimens, although the oocyst is the most important in transmission.

Prevalence

In the UK *Cyptosporidium* has been identified as the causative organism in 1%–5% of patients with acute diarrhoea, making it one of the most common identifiable causes of gastroenteritis in this country (Casemore and Jackson 1983; Hart et al. 1984; Hunt et al. 1984). Similar findings have been reported in North America and in other European countries (Jokipii et al. 1983; Holten-Andersen et al. 1984; Ratnam et al. 1985; Holley and Dover 1986). There is as yet no evidence to suggest that there is a reservoir of *Cryptosporidium* in the healthy homosexual population. L. Jokipii et al. (1985b) did not identify *Cryptosporidium* in the stools of any of 153 homosexual volunteers screened. McMillan and McNeillage (1984) failed to find *Crypto-*

sporidium in any of 85 homosexual men screened at a STD clinic in Edinburgh. Sixty-five homosexual men attending a STD clinic in London were screened for *Crypto-sporidium* and all were found to be negative, despite being selected for screening because other enteric protozoal organisms had been identified in their stools (E. Allason-Jones and P. G. Sargeaunt, 1986, unpublished work).

Transmission

With the increasing number of cases of *Cryptosporidium* reported, particularly in urban areas, the importance of animals as a reservoir of infection is diminishing. Inadequate treatment of domestic water supplies in some European countries has been implicated (Wolfson et al. 1984; L. Jokipii et al. 1985a) and person-to-person transmission has been documented (Baxby et al. 1983; Collier et al. 1984; Koch et al. 1985). There is currently no evidence that the organism can be sexually transmitted, although this remains a possibility. The source of infection in AIDS patients is obscure, but may be the same environmental sources that provide a reservoir of infection for the general population.

Clinical Features

The majority of people infected with *Cryptosporidium* probably develop symptoms. *Cryptosporidium* in immunocompetent patients presents as a flu-like illness with fever, malaise, nausea, watery diarrhoea and abdominal cramp. The incubation period is usually 5–12 days and symptoms last for 3–14 days. No specific treatment is available, but with supportive therapy by way of fluid and electrolyte replacement patients make a complete recovery. The organism may continue to be excreted for up to two weeks after the diarrhoea has resolved but there is no evidence as yet of long-term asymptomatic carriers (Current et al. 1983; Jokipii et al. 1983; Hart et al. 1984; A. M. M. Jokipii et al. 1985).

In patients with AIDS, the organism causes severe protracted diarrhoea. Fluid loss can be as much as 12 litres per day and the diarrhoea may persist for months, causing severe morbidity. Many AIDS patients remain infected up to the time they die. Whilst cryptosporidiosis is not directly responsible for the patient's death it may be a contributory factor (Centers for Disease Control 1982; Current et al. 1983; Pitlik et al. 1983; Navin and Juranek 1984).

Pathogenesis

The pathogenesis of cryptosporidiosis is not clear. The organism does not appear to invade epithelial cells, but either remains on the cell surface or is engulfed by cell wall membrane. The jejunum is the most severely affected area but the organism can be found throughout the intestinal tract. Histological changes are generally mild and non-specific, with shortened and distorted villi, lengthened hyperplastic crypts and a mild to moderate inflammatory cell infiltrate in the lamina propria and epithelium. Whether the diarrhoea is a defect of absorption or secretion is not known (Current et al. 1983; Casemore et al. 1985).

Diagnosis

Prior to 1978, diagnosis was made on biopsy specimens but, with improved staining techniques, stool specimens are all that is necessary in most cases (Garcia et al. 1983). A number of staining methods have been used, the most useful being the modified Ziehl-Neelsen stain or the phenol auramine fluorescence stain. The latter method works reliably even when faeces have been fixed in glutaraldehyde, making specimens from HIV-infected patients less hazardous to handle in the laboratory (Williams et al. 1985).

Treatment

No reliably effective treatment exists for cryptosporidiosis. In immunosuppressed patients whose immune status could be restored to normal, e.g. by stopping immunosuppressive drugs, spontaneous recovery has taken place. In patients with AIDS or other conditions in which the immune system cannot be reconstituted, a variety of drugs effective against other coccidian parasites or used in treating other protozoan infections in humans have been tried without success. The most useful drug used in the treatment of AIDS patients with cryptosporidiosis is spiramycin, a macrolide antibiotic that can be given orally and is rarely associated with serious side-effects. Treatment usually consists of 1 g spiramycin six to eight hourly, given for two weeks in the first instance. However, symptomatic improvement is not always accompanied by parasitological cure. A combination of oral quinine and clindamycin has been reported to be successful in a few cases, although this treatment has been associated with unpleasant side-effects (Centers for Disease Control 1984; Portnoy et al. 1984).

Isospora belli

Isospora belli is a coccidial protozoan parasite widespread in the animal kingdom and recognised as an unusual cause of gastrointestinal disease in man. Most reported cases have occurred in tropical areas and, in the immune-competent host, resulted in fever, headache, diarrhoea and abdominal pain (Matsubayashi and Nozawa 1948; Henderson et al. 1963). The disease is usually self-limiting but can run a protracted course of many months or even years and malabsorption has developed in some patients (Brandborg et al. 1970; Trier et al. 1974).

A number of cases of *I. belli* infection have been reported in immunocompromised patients, including more recently patients with AIDS. Infection in these patients has been associated with persistent, severe, watery diarrhoea, anorexia, weight loss and intermittent fever. The source of the organism in these individuals is unknown (Whiteside et al. 1984; De Hovitz et al. 1986).

Infection results from ingesting oocysts that excyst in the proximal small bowel. The life-cycle is similar to that of *Cryptosporidium*, with an asexual and sexual phase taking place but *within* the epithelial cells of the small intestine. Diagnosis can be made by finding oocysts in the faeces, but, as these structures are often scanty, examination of jejunal fluid or small intestinal biopsy material may be necessary. As well as demonstrating various endogenous stages of the organism within the enterocytes, histological examination may show the non-specific changes associated with *I. belli* infection, namely villous atrophy, elongated crypts, and a mild to moderate inflammatory cell infiltrate of the lamina propria (Brandborg et al. 1970; Trier et al. 1974).

Parasitological and symptomatic cure has been reported with pyrimethamine–sulphadiazine, trimethoprim–sulphamethoxazole, and furazolidone. Relapses appear to be common in immunosuppressed patients (Whiteside et al. 1984; De Hovitz et al. 1986).

Microsporidia

Microsporidia are intracellular spore-forming protozoa that infect a wide range of vertebrate and invertebrate species. Spores are excreted in the urine of infected animals and presumably ingested by other animals. Each spore characteristically contains a coiled polar filament, thought to be involved in injecting the sporoplasm into the host cell.

There are very few reports of microsporidiosis in man but meningoencephalitis, corneal infection and disseminated disease have been described (Wolf and Cowen 1937; Matsubayashi et al. 1959; Ashton and Wirasinha 1973; Margileth et al. 1973; Pinnolis et al. 1981). Recently, microsporidiosis has been reported in a small number of patients with AIDS. Five patients with persistent, unexplained diarrhoea were shown by electron microscopy to have *Microsporidia* in intestinal epithelial cells (Dobbins and Weinstein 1985; Modigliani et al. 1985; G. Nichols, 1986, personal communication). Whether the *Microsporidia* were involved in the pathogenesis of the diarrhoea is difficult to determine. One further AIDS patient has been described with myositis and spores of *Microsporidia* were found in muscle biopsies (Ledford et al. 1985).

The two genera of *Microsporidia* known to infect man are *Nosema* and *Encephalitozoon*. Despite the rare reports of encephalitozoonosis in man, antibodies to *E. cuniculi* were found in 12% of unselected Swedish residents returning from the tropics, suggesting infection in man is more common than previously thought (Parasitic Diseases Surveillance 1983). In a study of 30 Swedish homosexual men, two with AIDS and 28 with AIDS-related complex, *E. cuniculi* antibodies were found in 33% (Bergquist et al. 1984). All the infected men had visited the tropics at sometime; it is possible their sexual activities had increased the likelihood of their acquiring infection. No connection was claimed between the presence of *E. cuniculi* antibodies and patients' symptoms.

Diarrhoea in the Acquired Immune Deficiency Syndrome

Diarrhoea is a common complaint in patients with AIDS or AIDS-related complex. Despite careful investigation often no infective or malignant cause is identified; in these cases, supportive treatment by way of fluid replacement and anti-diarrhoeal agents is all that can be offered. Some episodes of diarrhoea may be due to enteric organisms that are currently considered to be non-pathogenic. Among the protozoal species *Blastocystis hominis* and the non-pathogenic amoebic species such as *Entamoeba hartmanni* have already been put forward as possible aetiological agents (Ricci et al. 1984; Rolston et al. 1986), but there is currently no evidence to support these suggestions. It is likely that organisms as yet unidentified are responsible for some cases of diarrhoea in immune-suppressed patients. Other mechanisms, for example immune-mediated tissue damage, may be involved in the initiation or perpetuation of diarrhoea in AIDS patients.

References

Allason-Jones E, Mindel A, Sargeaunt P, Williams P (1986) *Entamoeba histolytica* as a commensal intestinal parasite in homosexual men. N Engl J Med 315: 353–356

Angus KW (1983) Cryptosporidiosis in man, domestic animals and birds: a review. J R Soc Med 76: 62–70

Ashton N, Wirasinha PA (1973) Encephalitozoonosis (nosematosis) of the cornea. Br J Opthalmol 57: 669–674

Baxby D, Hart CA, Taylor C (1983) Human cryptosporidiosis: a possible case of hospital cross infection. Br Med J 287: 1760–1761

Bergquist R, Morfeldt-Mansson L, Pehrson PO, Petrini B, Wasserman J (1984) Antibody against *Encephalitozoon cuniculi* in Swedish homosexual men. Scand J Infect Dis 16: 389–391

Bienzle U, Coester CH, Knobloch J, Guggenmoos-Holzmann I (1984) Protozoal enteric infections in homosexual men. Klin Wochenschr 62: 323–327

Brandborg LL, Goldberg SB, Breidenbach WC (1970) Human coccidiosis – a possible cause of malabsorption. N Engl J Med 283: 1306–1313

Burnham WR, Reeve RS, Finch RG (1980) *Entamoeba histolytica* infection in male homosexuals. Gut 21: 1097–1099

Casemore DP, Jackson B (1983) Sporadic cryptosporidiosis in children. Lancet ii: 679

Casemore DP, Sands RL, Curry A (1985) *Cryptosporidium* species, a "new" human pathogen. J Clin Pathol 38: 1321–1336

Centers for Disease Control (1982) Cryptosporidiosis: assessment of chemotherapy of males with acquired immune deficiency syndrome (AIDS). MMWR 31: 589–592

Centers for Disease Control (1984) Update: treatment of cryptosporidiosis in patients with acquired immune deficiency syndrome (AIDS). MMWR 33: 117–119

Chin ATL, Gerken A (1984) Carriage of intestinal protozoal cysts in homosexuals. Br J Vener Dis 60: 193–195

Collier AC, Miller RA, Meyers JD (1984) Cryptosporidiosis after marrow transplantation: person-to-person transmission and treatment with spiramycin. Ann Intern Med 101: 205–206

Current WL, Reese NC, Ernst JV, Bailey WS, Heyman MB, Weinstein WM (1983) Human cryptosporidiosis in immunocompetent and immunodeficient persons. N Engl J Med 308: 1252–1257

De Hovitz JA, Pape JW, Boncy M, Johnson WD (1986) Clinical manifestations and therapy of *Isospora belli* infection in patients with the acquired immune deficiency syndrome. N Engl J Med 315: 87–90

Dobbins WO, Weinstein WM (1985) Electron microscopy of the intestine and rectum in acquired immunodeficiency syndrome. Gastroenterology 88: 738–749

Garcia LS, Bruckner DA, Brewer TC, Shimizu RY (1983) Techniques for the recovery and identification of *Cryptosporidium* oocysts from stool specimens. J Clin Microbiol 18: 185–190

Gathiram V, Jackson TFHG (1985) Frequency distribution of *Entamoeba histolytica* zymodemes in a rural South African population. Lancet i: 719–721

Goldmeier D, Sargeaunt PG, Price AB et al. (1986) Is *Entamoeba histolytica* in homosexual men a pathogen? Lancet i: 641–644

Green EL, Miles MA, Warhurst DC (1985) Immunodiagnostic detection of *Giardia* antigen in faeces by a rapid visual enzyme-linked immunosorbent assay. Lancet ii: 691–693

Hakansson C, Thoren K, Norkrans G, Johannisson G (1984) Intestinal parasitic infection and other sexually transmitted diseases in asymptomatic homosexual men. Scand J Infect Dis 16: 199–202

Hart CA, Baxby D, Blundell N (1984) Gastroenteritis due to *Cryptosporidium*: a prospective survey in a children's hospital. J Infect 9: 264–270

Henderson HE, Gillepsie GW, Kaplan P, Steber M (1963) The human *Isospora*. Am J Hyg 78: 302–309

Holley HP, Dover C (1986) *Cryptosporidium*: a common cause of parasitic diarrhoea in otherwise healthy individuals. J Infect Dis 153: 365–367

Holten-Andersen W, Gerstoft J, Henriksen SA, Strandberg Pedersen N (1984) Prevalence of *Cryptosporidium* among patients with acute enteric infection. J Infect 9: 277–282

Hunt DA, Shannon R, Palmer SR, Jephcott AE (1984) Cryptosporidiosis in an urban community. Br Med J 289: 814–816

Hurwitz AL, Owen RL (1978) Veneral transmission of intestinal parasites. (Medical Information). West J Med 128: 89–91

Jackson TFHG, Sargeaunt PG, Williams JE, Simjee AE (1982) Observations on zymodeme studies of *Entamoeba histolytica* in Durban, South Africa. Arch Invest Med (Mex) 13 (suppl. 3): 83–88

Jackson TFHG, Gathiram V, Simjee AE (1985) Seroepidemiological study of antibody responses to the zymodemes of *Entamoeba histolytica*. Lancet i: 716–719

Jokipii AMM, Hemila M, Jokipii L (1985) Prospective study of acquisition of *Cryptosporidium*, *Giardia lamblia* and gastrointestinal illness. Lancet ii: 487–489

Jokipii L, Pohjola S, Jokipii AMM (1983) *Cryptosporidium*: a frequent finding in patients with gastrointestinal symptoms. Lancet ii: 358–360

Jokipii L, Pohjola S, Jokipii AMM (1985a) Cryptosporidiosis associated with traveling and giardiasis. Gastroenterology 89: 838–842

Jokipii L, Pohjola S, Valle SL, Jokipii AMM (1985b) Frequency, multiplicity and repertoire of intestinal protozoa in healthy homosexual men and in patients with gastrointestinal symptoms. Ann Clin Res 17: 57–59

Kazal HL, Sohn N, Carrasco JI, Robilotti JG, Delaney WE (1976) The gay bowel syndrome: clinicopathologic correlation in 260 cases. Ann Clin Lab Sci 6: 184–192

Kean BH (1976) Venereal amebiasis. NY State J Med 6: 930–931

Kean BH, William DC, Luminais SK (1979) Epidemic of amoebiasis and giardiasis in a biased population Br J Vener Dis 55: 375–378

Keystone JS, Keystone DL, Proctor EM (1980) Intestinal parasitic infections in homosexual men: prevalence, symptoms and factors in transmission. Can Med Assoc J 123: 512–514

Koch KL, Phillips DJ, Aber RC, Current WL (1985) Cryptosporidiosis in hospital personnel: evidence for person to person transmission. Ann Intern Med 102: 593–596

Ledford DK, Overman MD, Gonzalvo A, Cali A, Mester SW, Lockey RF (1985) Microsporidiosis myositis in a patient with the acquired immune deficiency syndrome. Ann Intern Med 102: 628–630

Margileth AM, Strano AJ, Chandra R, Neafic R, Blum M, McCully RM (1973) Disseminated nosematosis in an immunologically compromised infant. Arch Pathol 95: 145–150

Markell EK, Havens RF, Kuritsubo RA, Wingerd J (1984) Intestinal protozoa in homosexual men of the San Francisco Bay area: prevalence and correlates of infection. Am J Trop Med Hyg 33: 239–245

Mathews HM, Moss DM, Healy GR, Mildvan D (1986) Isoenzyme analysis of *Entamoeba histolytica* isolated from homosexual men. J Infect Dis 153: 793–795

Matsubayashi H, Nozawa T (1948) Experimental infection of *Isospora hominis* in man. Am J Trop Med Hyg 28: 633–637

Matsubayashi H, Koike T, Mikata I, Takei H, Hagiwara S (1959) A case of *Encephalitozoon*-like body infection in man. Arch Pathol 67: 181–187

McMillan A, McNeillage GJC (1984) Comparison of the sensitivity of microscopy and culture in the laboratory diagnosis of intestinal protozoal infection. J Clin Pathol 37: 809–811

McMillan A, Gilmour HM, McNeillage G, Scott GR (1984) Amoebiasis in homosexual men. Gut 25: 356–360

Meyers JD, Kuharic HA, Holmes KK (1977) *Giardia lamblia* infection in homosexual men. Br J Ven Dis 53: 54–55

Mildvan D, Gelb AM, William D (1977) Venereal transmission of enteric pathogens in male homosexuals. JAMA 238: 1387–1389

Modigliani R, Bories C, Le Charpentier Y, Salmeron M, Messing B, Galian A, Rambaud JC, Lavergne A, Cochand-Priollet B, Desportes I (1985) Diarrhoea and malabsorption in acquired immune deficiency syndrome: a study of four cases with special emphasis on opportunistic protozoan infestations. Gut 26: 179–187

Most H (1968) Manhattan: "A tropic isle?" Am J Trop Med Hyg 17: 333–354

Navin TR, Juranek DD (1984) Cryptosporidiosis: clinical, epidemiologic and parasitologic review. Rev Infect Dis 6: 313–327

Nime FA, Burek JD, Page DL, Holscher MA, Yardley JH (1976) Acute enterocolitis in a human being infected with the protozoan *Cryptosporidium*. Gastroenterology 70: 592–598

Parasitic Diseases Surveillance (1983) Antibody to *Encephalitozoon cuniculii* in man. WHO Weekly Epidem Rec 58: 30–32

Pearce RB (1983) Intestinal protozoal infections and AIDS (Letter). Lancet ii: 51

Phillips SC, Mildvan D, William DC, Gelb AM, White MC (1981) Sexual transmission of enteric protozoa and helminths in a venereal-disease-clinic population. N Engl J Med 305: 603–606

Pinnolis M, Egbert PR, Font RL, Winter FC (1981) Nosematosis of the cornea. Case report, including electron microscopic studies. Arch Ophthalmol 99: 1044–1047

Pitlik SD, Fainstein V, Garza D et al. (1983) Human cryptosporidiosis: spectrum of disease. Report of six cases and review of the literature. Arch Intern Med 143: 2269–2275

Pohlenz J, Moon HW, Cheville NF, Bemrick WJ (1978) Cryptosporidiosis as a probable factor in neonatal diarrhea of calves. J Am Vet Med Assoc 172: 452–457

Portnoy D, Whiteside ME, Buckley E, MacLeod CL (1984) Treatment of intestinal cryptosporidiosis with spiramycin. Ann Intern Med 101: 202–204

Quinn TC, Stamm WE, Goodell SE, Mkrtichian E, Benedetti J, Corey L, Schuffler MD, Holmes KK (1983) The polymicrobial origin of intestinal infections in homosexual men. N Engl J Med 309: 576–582

Ratnam S, Paddock J, McDonald E, Whitty D, Jong M, Cooper R (1985) Occurrence of *Cryptosporidium* oocysts in fecal samples submitted for routine microbiological examination. J Clin Microbiol 22: 402–404

Rendtorff RC (1954) The experimental transmission of human intestinal protozoan parasites II. *Giardia lamblia* cysts given in capsules. Am J Hyg 59: 209–220

Ricci N, Toma P, Furlani M, Caselli M, Gullini S (1984) *Blastocystis hominis*: a neglected cause of diarrhoea? (Letter). Lancet i: 966

Rolston KVI, Hoy J, Mansell PWA (1986) Diarrhoea caused by "non-pathogenic amoebae" in patients with AIDS. N Engl J Med 315: 192

Saltzberg DM, Hall-Craggs M (1986) Fulminant amebic colitis in a homosexual man. Am J Gastroenterol 81: 209–212

Sargeaunt PG (1985) Zymodemes expressing possible genetic exchange in *Entamoeba histolytica*. Trans R Soc Trop Med Hyg 79: 86–89

Sargeaunt PG, Williams JE (1978) Electrophoretic isoenzyme patterns of *Entamoeba histolytica* and *Entamoeba coli*. Trans R Soc Trop Med Hyg 72: 164–166

Sargeaunt PG, Williams JE, Grene JD (1978) The differentiation of invasive and non-invasive *Entamoeba histolytica* by isoenzyme electrophoresis. Trans R Soc Trop Med Hyg 72: 519–521

Sargeaunt PG, Williams JE, Kumate J, Jimenez E (1980) The epidemiology of *Entamoeba histolytica* in Mexico City: a pilot survey. I. Trans R Soc Trop Med Hyg 74: 653–656

Sargeaunt PG, Jackson TFHG, Simjee A (1982a) Biochemical homogeneity of *Entamoeba histolytica* isolates, especially those from liver abscess. Lancet i: 1386–1388

Sargeaunt PG, Williams JE, Bhojnani R, Campos JE, Gomez A (1982b) The epidemiology of *Entamoeba histolytica* in a rural and an urban area of Mexico: a pilot survey. II. Trans R Soc Trop Med Hyg 76: 208–210

Sargeaunt PG, Williams JE, Bhojnani R, Kumate J, Jimenez E (1982c) A review of isoenzyme characterization of *Entamoeba histolytica* with particular reference to pathogenic and non-pathogenic stocks isolated in Mexico. Arch Invest Med (Mex). 13 (Suppl 3): 89–94

Sargeaunt PG, Williams JE, Jackson TFHG, Simjee AE (1982d) A zymodeme study of *Entamoeba histolytica* in a group of South African school children. Trans R Soc Trop Med Hyg 76: 401–402

Sargeaunt PG, Oates JK, MacLennan I, Oriel JD, Goldmeier D (1983) *Entamoeba histolytica* in male homosexuals. Br J Vener Dis 59: 193–195

Sargeaunt PG, Baveja UK, Nanda R, Anand BS (1984) Influence of geographical factors in the distribution of pathogenic zymodemes of *Entamoeba histolytica*: identification of zymodeme XIV in India. Trans R Soc Trop Med Hyg 78: 96–101

Sargeaunt PG, Jackson TFHG, Wiffen S et al. (1987) The reliability of *Entamoeba histolytica* zymodemes in clinical laboratory diagnosis. Arch Invest Med (Mex) 18: 69–75

Schmerin MJ, Gelston A, Jones TC (1977) Amebiasis. An increasing problem among homosexuals in New York City. JAMA 238: 1386–1387

Schmerin MJ, Jones TC, Klein H (1978) Giardiasis: association with homosexuality. Ann Intern Med 88: 801–803

Sorvillo FJ, Strassburg MA, Seidel J et al. (1986) Amebic infections in asymptomatic homosexual men, lack of evidence of invasive disease. Am J Public Health 76: 1137–1139

Trier JS, Moxey PC, Schimmel EM, Robles E (1974) Chronic intestinal coccidiosis in man: intestinal morphology and response to treatment. Gastroenterology 66: 923–935

Tyzzer EE (1907) A sporozoan found in the peptic glands of the common mouse. Proc Soc Exp Biol Med 5: 12–13

Whiteside ME, Barkin JS, May RG, Weiss SD, Fischl MA, MacLeod CL (1984) Enteric coccidiosis among patients with the acquired immune deficiency syndrome. Am J Trop Med Hyg 33: 1065–1072

William DC, Shookhoff HB, Felman YM, DeRamos SW (1978) High rates of enteric protozoal infections in selected homosexual men attending a venereal disease clinic. Sex Transm Dis 5: 155–157

Williams JE, Ellis DS, Smith MD, Daziel R (1985) Safe method for identifying cryptosporidium cysts in the faeces of patients with suspected AIDS or those infected with other serious concomitant pathogens (Letter). J Clin Pathol 38: 1313–1314

Wolf A, Cowen D (1937) Granulomatous encephalomyelitis due to an *Encephalitozoon* (encephalitozoic encephalomyelitis). Bull Neurol Inst NY 6: 306–371

Wolfson JS, Hopkins CC, Weber DJ, Richter JM, Waldron MA, McCarthy DM (1984) An association between *Cryptosporidium* and *Giardia* in stool (Letter). N Engl J Med 310: 788

Ylvisaker JT, McDonald GB (1980) Sexually acquired amebic colitis and liver abscess. West J Med 132: 153–157

Chapter 5

Hepatitis

I. V. D. Weller

Introduction

This chapter is largely devoted to hepatitis B virus infection, because of its high prevalence amongst homosexual men and the morbidity and mortality associated with
the chronic carrier state, which develops in approximately 5% of those infected.
There are three other main categories of viral hepatitis (Table 5.1).

Hepatitis A is caused by a small RNA virus (HAV). It is spread by the faecal/oral
route and excreted in the stool for up to two weeks before the onset of symptoms.
With increasing standards of hygiene in the developed world, an older section of the
population has become susceptible. Fewer than a fifth of young heterosexual adults
in London are immune. In 1980 a study showed that homosexual men attending a
sexually transmitted disease (STD) clinic in Seattle had an almost three-fold higher
prevalence of anti-HAV compared to heterosexual controls. The acquisition of infection was correlated with frequent oral/anal contact and was added to the list of enteric
infections which can be sexually transmitted amongst homosexual men (Corey and
Holmes 1980). Similar findings were then reported in the UK, the Netherlands and
Sweden (Mindel and Tedder 1981; Coutinho et al. 1983; Christenson et al. 1982).
Recovery is the rule following hepatitis A virus infection, although prolonged cholestasis with a more prolonged course and relapse have been described (Teixeira et al.
1982).

Table 5.1. Hepatitis viruses in homosexual men

Common	Rare
Hepatitis B virus	NANB
Hepatitis A virus	Delta
	CMV
	EBV

NANB, non-A, non-B; CMV, cytomegalovirus;
EBV, Epstein–Barr virus.

Non-A, non-B viruses (hepatitis C) emerged as important causes of hepatitis once serological tests were available to identify HAV, HBV, cytomegalovirus (CMV) and Epstein–Barr virus (EBV). CMV and EBV occasionally cause a hepatitis as part of a glandular fever syndrome. There is no convincing evidence that non-A, non-B viruses are sexually transmitted, but there are reports of possible transmission with close personal contact. This is difficult to test, because there is as yet no serological marker for infection. There may be two agents transmitted via percutaneous routes, blood and blood products, so that intravenous drug users (Weller et al. 1984) and the recipients of such products (such as haemophiliacs) are at greatest risk. It would seem that the majority of infected individuals may develop chronic hepatitis. Non-A, non-B viruses account for 15%–30% of sporadic acute hepatitis and chronicity is unusual. Non-A, non-B hepatitis spread by the oral/faecal route has caused epidemics in Africa and Asia, with an incubation period similar to that of hepatitis A, and there appears to be no carrier state (Iwarson 1987).

Hepatitis delta virus (HDV) is a defective RNA virus which requires the presence of hepatitis B to produce infection and disease (Rizzetto and Verme 1985). In serum, delta antigen and a small RNA molecule are hidden in a subpopulation of particles of hepatitis B surface antigen (HBsAg) (Fig. 5.1). It occurs largely as a superinfection in carriers of hepatitis B virus and may cause a transient infection, severe acute or fulminant hepatitis, or worsened chronic liver disease. It is uncommon in homosexual men (Weller et al. 1983b) and in the developed world intravenous drug users appear to be at greatest risk.

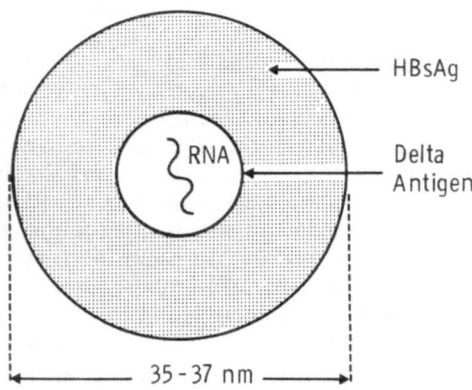

Fig. 5.1. Delta virus in serum. *HBsAg*, hepatitis B serum antigen.

Diagnosis of Acute Hepatitis

There are no major differences in the clinical features of acute viral hepatitis caused by these viruses. More than half of acute infections are subclinical. The clinical illness begins with non-specific symptoms such as fever, headache and fatigue, and jaundice follows. Symptoms and signs of other concurrent sexually transmitted diseases should always be looked for and it is important to enquire about, and see, contacts.

A history of travel, intravenous drug use, tattoos, recent transfusion or percutaneous exposure may provide clues as to the likely viral cause. Hepatotoxins such as alcohol and drugs should also be excluded as a cause of hepatitis. Routine liver function tests are unhelpful in identifying the responsible virus; they may help to distinguish between hepatitis and cholestasis due to other extra-hepatitic or intra-hepatitic liver disease. However, prolonged cholestasis does occur with viral hepatitis. Tests of synthetic function such as prothrombin time and serum albumin are useful in determining the severity of the hepatitis.

Specific serological tests are required to differentiate the viral infections. HAV, EBV and CMV infections are best diagnosed by tests for specific serum IgM antibodies which persist for a short time and indicate recent infection. Acute HBV infection is best diagnosed by the presence of IgM antibodies to the hepatitis B core antigen (anti-HBc) (see Fig. 5.5), since HBsAg may disappear from serum before presentation. The presence of this antibody will also differentiate acute HBV infection from other causes of acute hepatitis in a chronic HBV carrier, such as concomitant HDV infection. Conventional assays for delta antigen and antibodies are now available. There are still no specific serological tests for non-A, non-B viruses and the diagnosis is one of exclusion of HAV, HBV, HDV, CMV, EBV, alcohol, drugs and other causes of chronic hepatitis which may occasionally present acutely, namely rarely autoimmune chronic active hepatitis and Wilson's disease.

Management of Acute Hepatitis

Our knowledge of the spectrum of hepatotrophic viruses and their structure, molecular biology, pathogenic effects and the immune response of the host have increased enormously. However, these advances have had, as yet, little influence on management. Acute hepatitis is usually self-limiting and management largely supportive. Most patients can be managed as outpatients but acute liver failure and its complications demand urgent admission to an intensive care unit. Fulminant hepatitis is rare, occurring in much fewer than 1 in a 100 cases of acute hepatitis.

For the uncomplicated case, a low-fat, high-calorie diet has been advocated, probably because patients find it more palatable rather than because of evidence that it is beneficial. Patients should be encouraged to eat what they want. Similarly the patient and his illness will determine to a large extent the level of physical activity allowed. A number of studies in American military personnel have shown that strict bed rest confers no advantage over graded exercise (Swift et al. 1950; Chalmers et al. 1955; Repsher and Freebern 1969). However, these apply to the convalescent phase of acute hepatitis, probably hepatitis A virus infection and in young previously healthy men. It would seem wise to prescribe bed rest for the patient in the early phase of the illness, with no strenuous exercise until the convalescent phase is well established. Alcohol is usually excluded on the grounds of its hepatotoxicity and the effect of the illness itself will often preclude its use. The length of abstention is arbitrary but it would seem wise to exclude it at least until liver function tests are normal. Surgery in acute hepatitis carries a high mortality and morbidity (Harville and Summerskill 1963) and is unnecessary now that non-invasive methods for the diagnosis of extra-hepatitic jaundice are available.

Specific therapies are not indicated in an often mild and self-limiting viral illness but have been used to treat severe acute and fulminant hepatitis. Several controlled trials of corticosteroids have shown no benefit, and, indeed, in one there was a high mortality in the treatment group (Ware et al. 1981). It has been suggested that the incidence of chronic liver disease may be increased by immunosuppression.

Hepatitis B Virus Infection

The Structure of Hepatitis B Virus

Since the discovery of the Australia antigen (Blumberg et al. 1965), and without a tissue culture system for the study of the virus, it is remarkable how much we now know about the structure and biology of HBV. HBsAg is a protein with non-glycosylated and glycosylated forms. It circulates in serum of an infected individual in the form of small 22 nm spherical lipoprotein particles and occasional filaments. HBsAg is produced in vast excess of its covering of the complete virion, or Dane particle, which is 42 nm in size (Dane et al. 1970) (Fig. 5.2). HBsAg does demonstrate antigenic heterogeneity, but there is a common group-specific antigen a and two mutually exclusive major subdeterminants, d/y or w/r. There is a marked geographical variation in the prevalence of different subtypes but the ad subtype is commonest in homosexual men in the developed world.

Disruption of the HBsAg envelope covering the 42 nm particle by detergent reveals a nucleocapsid. It is not normally detected in the serum of an infected individual and it is antigenically distinct from HBsAg. These particles contain a partially double-

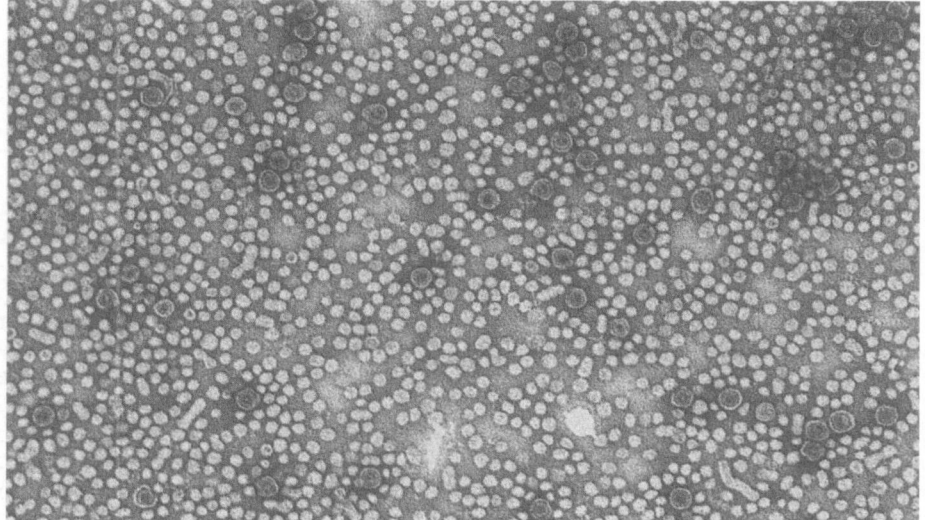

Fig. 5.2. Electron micrograph of an HBV carrier serum showing 42 mm Dane particles and smaller 22nm spherical forms and occasional filaments. ×44800.

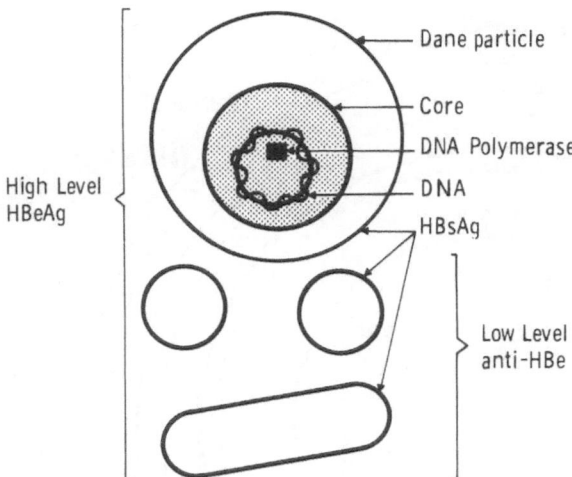

Fig. 5.3. Markers of hepatitis B viral replication. *HBsAg*, hepatitis B serum antigen (*Ag*).

stranded circular DNA molecule and the enzyme DNA polymerase (Fig. 5.3). In vitro, this enzyme completes the gap in the short strand of the DNA. A third antigen (HBeAg) distinct from HBsAg and its subdeterminants was discovered by immunodiffusion (Magnius and Epsmark 1972). Because of its association with the presence of complete virus particles in the serum and infectivity, it was felt that there must be a close association between HBe antigen and HBc antigen. This was shown when HBe antigen was released by proteolytic digestion of HBc antigen cloned in *Escherichia coli* (Mackay et al. 1981); HBe antigen, like HBs antigen, is produced in excess of what the virion's nucleocapsid requires and provides a useful marker of infectivity. The core particle of HBV contains a DNA molecule which is partially double-stranded, and the single-strand region occupies 15%–50% of the overall length in different molecules. The DNA therefore consists of the long strand of constant length (3200 nucleotides) and the short strand, which varies in length from 1700 to 2800 nucleotides in different molecules. The nucleotide sequencing of cloned HBV DNA by restriction endonuclease analysis has identified four regions of the genome that code without interruption for polypeptides (Fig. 5.4). The S region codes for HBsAg, the C region for HB core antigen, and the P region, which is the largest and covers more than 75% of the genome, codes for the DNA polymerase. The fourth region, X, is small and its function is unknown (Tiollais et al. 1981).

Natural History of Acute and Chronic HBV Infection

Sexual transmission is an important mode of transmission in the developed world and it occurs in heterosexuals as well as in male homosexuals (Perillo et al. 1979; Heathcote and Sherlock 1973; Scott 1980; Schreeder et al. 1982; Reiner et al. 1982). In the Third World, however, with high HBV carrier rates, perinatal infection and infection in early childhood, presumably resulting from close physical contact with carrier parents or relatives and other children, are more important modes of transmission.

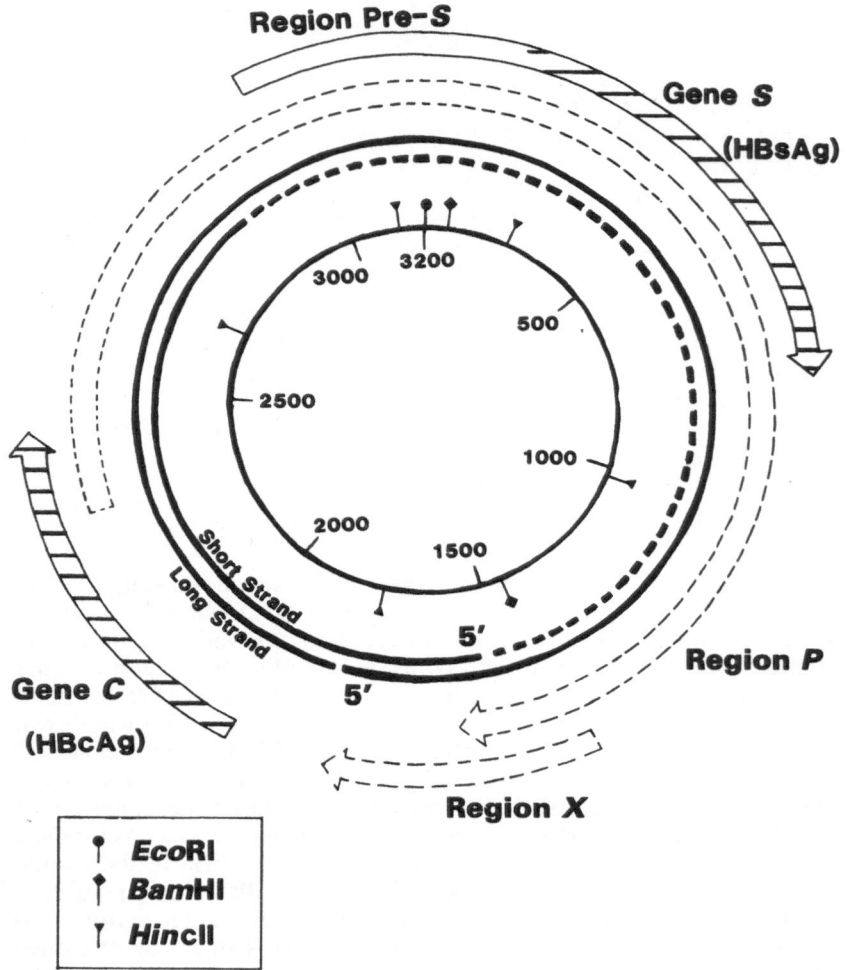

Fig. 5.4. The HBV genome (modified from Tiollais et al. 1981). *HBsAg*, hepatitis B serum antigen (*Ag*).

Infection with HBV is followed by a long incubation period of 35–100 days before the clinical illness. Serological events following exposure, acute hepatitis and recovery are well characterised, with the appearance of antibodies to the three main antigens previously discussed – anti-HBc, anti-HBe and anti-HBs (Fig. 5.5). Complete viral replication marked by the presence of HBe antigen and DNA polymerase and serum HBV DNA is at its peak before the peak in liver enzymes. This, together with the fact that a large number of chronic carriers with high levels of HBV replication have normal transaminases, suggests that hepatocyte damage is caused more by the host immune response to the virus or virally determined antigens than by the virus itself. These markers of hepatitis B virus replication disappear rapidly in the early stages of the acute illness, followed by the development of anti-HBs if elimination of the infection is to occur. The first antibody to appear is anti-HBc, which appears early

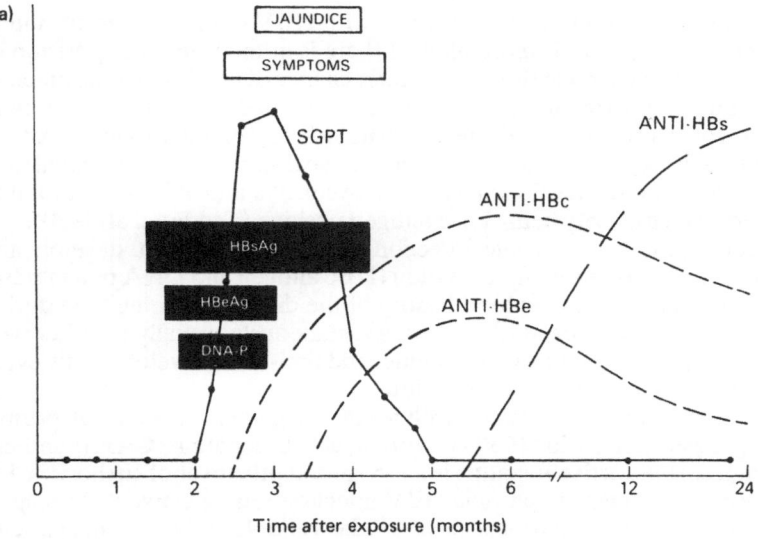

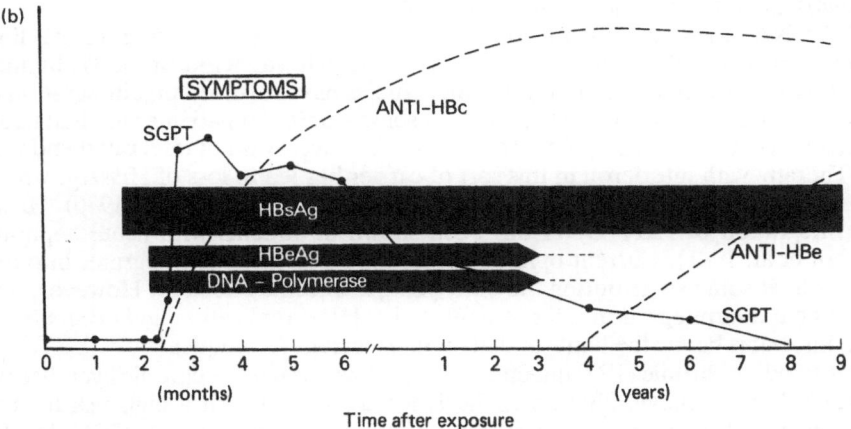

Fig. 5.5 a Serological events during acute infection with recovery; **b** serological events during acute to chronic infection (from Hoofnagle 1980). *DNA-P*, DNA polymerase; *SGPT*, serum glutamic oxaloacetic transaminase; *Ag*, antigen.

in the infection and is always present by the time of the clinical illness. There may be a gap in the time between the disappearance of HBsAg and appearance of anti-HBs (the serological window) (Hoofnagle 1980). However, newer assays for HBsAg and anti-HBs using monoclonal reagents shorten this window (Fagan and Williams 1986). The time of clearance of HBsAg is variable. Approximately 5% of patients clear HBsAg by the onset of clinical symptoms, but in the main by four months from the illness virtually all those that are going to clear HBsAg do so. Both anti-HBs and anti-HBc are long-lasting antibodies. Anti-HBs is the major protective antibody, as shown by the trials of formalin-inactivated HBsAg vaccines.

Between 5% and 10% of individuals infected with HBV develop a chronic infection (Nielsen et al. 1971). This is more likely if there is immaturity or depression of the immune system. For example there is a greater chance of persistent antigenaemia in the neonate infected perinatally, in patients on renal dialysis, in those with lymphomas and in Down's syndrome. There are 15 million to 200 million carriers of the hepatitis B virus in the world and, in seroepidemiological surveys of homosexual men attending STD clinics in the developed world, between 3% and 5% will be found to be HBV carriers. The majority being HBe antigen positive (Tedder et al. 1980).

With the development of chronic infection, anti-HBs does not develop and HBsAg and the markers of the complete virion (HBe antigen and DNA polymerase) persist (Fig. 5.5). As part of the natural history of the disease, at some time during the chronic infection HBe antigen disappears spontaneously, usually together with the complete virus particle and its components, and anti-HBe develops. This event will be discussed under antiviral therapy, below.

The chronic carrier state is associated with a wide range of liver injury. A normal liver, chronic persistent hepatitis (CPH), chronic active hepatitis (CAH), and cirrhosis. Although chronic active hepatitis is a recognised precurser of cirrhosis, CPH is usually regarded as benign. In chronic HBV infection retrospective and prospective studies have shown that the prognosis of CPH is variable and that some patients do progress to CAH and cirrhosis, particularly if they remain HBe antigen positive (Chadwick et al. 1979; Aldershville et al. 1982).

HBV DNA was first shown to be integrated into a hepatoma tumour cell line (Chakraborty et al. 1980; Edman et al. 1980) and then it was demonstrated in human tumour tissue (Brechot et al. 1980). Further studies have revealed that integration is present in patients with chronic hepatitis. In some studies on patients who had been infected for only a few years, the HBV DNA in the liver was not integrated and was free. Therapy with interferon in this sort of carrier has led to loss of HBsAg, which again suggests that the HBV DNA was not integrated (Thomas et al. 1986). However, integration of HBV DNA has been shown in patients with acute hepatitis (Brechot et al. 1981). Current opinion is that the HBV genome integrates into the hepatocyte at some stage during the early years of chronic infection. However, it is still possible that integration is a prerequisite for HBV replication and occurs in all infections, but is below the limits of detection of current techniques.

Longstanding chronic HBV infection and the development of chronic liver disease ultimately but not inevitably into cirrhosis are associated with a high risk for the development of primary hepatocellular carcinoma. The role of HBV in the pathogenesis of hepatoma has been established in a number of ways. There is still some controversy about which is most important, integrated HBV DNA or cirrhosis, but there is a high incidence of hepatoma in areas with a high incidence of chronic HBV infection. There is a high incidence of serum HBV markers in patients with hepatoma and, where perinatal transmission is common, their mothers (Szmuness 1978). HBsAg was demonstrated in the tumour and non-tumour tissue of patients with hepatoma. Hepatoma tumour cell lines secrete HBsAg and hepatoma develops when these cells are injected into mice. There is integration of the HBV genome in such tumour cell lines, in liver cell DNA during chronic infection and in the tumour and surrounding non-tumour tissue of hepatoma in man. Finally there are hepatitis viruses structurally similar to HBV causing persistent infection and hepatocellular carcinoma in woodchucks (Summers 1981; Summers and Mason 1982).

Antiviral Therapy of Chronic Hepatitis B Virus Infection

The antiviral therapy of chronic hepatitis B virus infection began in the mid 1970s with human leukocyte interferon and adenine arabinoside (ARA). A pioneer in this field was Thomas Merigan, and his group described their six year experience (Scullard et al. 1981b). The majority of patients showed a transient inhibition of productive viral replication in response to one or other or both of these drugs during therapy (type I response). However, a minority underwent an HBeAG to anti-HBe seroconversion (type II response) and a few lost all markers of HBV replication (type III response). In patients who underwent HBeAg to anti-HBe seroconversion, and in those that lost all markers of HBV replication symptoms, liver function tests and liver histology improved. These studies, however, were not controlled, two drugs were used and it is difficult to decide which, if either, was responsible for these changes.

ARA-AMP is the monophosphate ester of ARA-A and 400 times more water soluble (Weller et al. 1982a). It can therefore be given by intramuscular injection. Early studies with this compound were promising: in five patients a short course produced a transient inhibition of HBV replication and, in three consecutive patients treated for one month, all underwent HBe to anti-HBe seroconversion following a marked rise in the level of aspartate transaminase (Fig. 5.6). This pattern of response had also been seen with other antiviral therapy and was attributed to a heightened immune response to virally infected cells leading to increased liver cell necrosis as a result of perhaps a cytotoxic T cell response and elimination of those cells productively replicating HBV.

In a controlled trial in which 15 patients were treated with ARA-AMP for one month, and 14 untreated patients were included as controls, four of the 15 treated patients underwent HBeAg to anti-HBe seroconversion (type II response). Eleven patients showed a type I response. There were no such changes in the control patients (Weller et al. 1985). Unfortunately the interpretation of these findings was complicated by the fact that three of the patients had intercurrent infection with other hepatotrophic viruses (HDV in two and HAV in one), which may have inhibited HBV replication (Viola et al. 1982). Nevertheless, this antivirally induced seroconversion was higher than that previously described to occur spontaneously in homosexual men.

As so often happens, therapy had moved on in advance of our understanding of the natural history of chronic HBV infection (Table 5.2) (Weller et al. 1986). Early in the chronic carrier state the HBV carrier supports productive viral replication. Complete virions are produced in the liver, released into the serum and detected by the presence of HBV DNA polymerase and HBV DNA. HBe antigen is also present in the serum. During the natural history of infection, spontaneous loss of HBeAg occurs with the development of anti-HBe. Associated with this there is a decrease in infectivity. During the HBeAg to anti-HBe transition there would appear to be a heightened cell-mediated immune response, with effector cells, probably cytotoxic T lymphocytes, directed at target viral antigens on or in hepatocytes, most likely core proteins (Eddleston et al. 1982; Pignatelli et al. 1987)). This results in the histological appearance during this phase of lobular hepatitis and is reflected by marked rises in levels of serum transaminases (Liaw et al. 1982). Following the loss of HBeAg and the development of anti-HBe, inflammatory activity in the liver settles and the levels of transaminases return to normal or near normal. The patient then enters a phase of inactive liver disease.

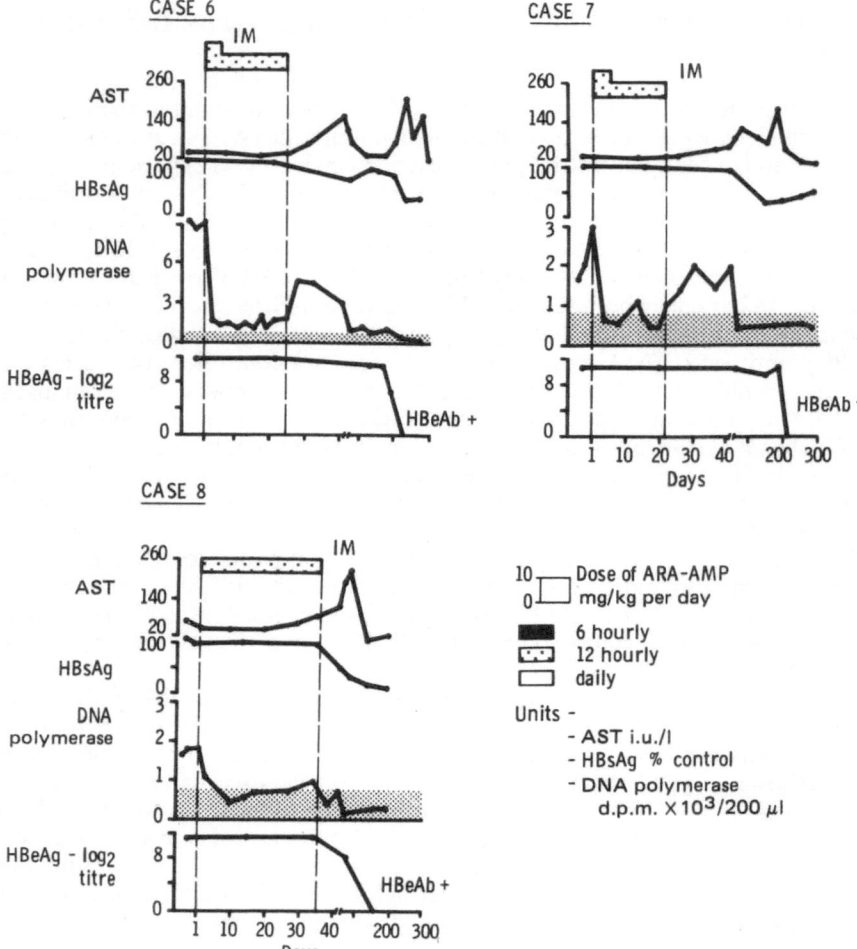

Fig. 5.6. Prolonged inhibition of viral replication with ARA-AMP (from Weller et al. 1982a). *AST*, aspartate transaminase; *IM* intramuscular; *Ag*, antigen; *Ab*, antibody.

If this transition occurs early in the natural history of infection and the switch is not prolonged, the patient is likely to have a chronic persistent hepatitis or very mild chronic active hepatitis. If this switch is prolonged or occurs late in the natural history, then the patient may well have developed active cirrhosis and inflammatory activity will settle, leaving an inactive cirrhosis.

Our fuller understanding of the natural history of chronic infection led to concern that some of the responses described with previous antiviral therapy may well have been spontaneous events. In particular many of the patients that were being treated were treated on liver units and therefore a pre-selection of patients with high levels of transaminases, i.e. those close to spontaneous seroconversion, might have occurred.

A recent review of 51 treatment groups in trials of antiviral agents in chronic hepatitis B virus infection revealed that over half had fewer than 10 patients treated

Table 5.2. Natural history of chronic HBV infection

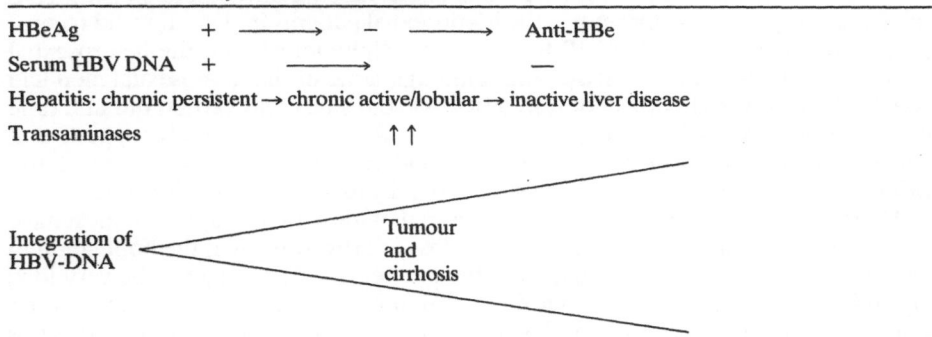

Early in the chronic carrier state, the HBV carrier supports productive viral replication. HBeAg and HBV-DNA are present in serum. During the natural history of chronic infection, loss of HBeAg and HBV-DNA occur, with the development of anti-HBe. During this transition there is increased hepatic inflammatory activity, with a chronic active or lobular hepatitis with raised transaminases. With the development of anti-HBe, inflammatory activity in the liver settles and there is inactive liver disease. The combination of increasing levels of integrated HBV-DNA in hepatocytes and cirrhosis leads to an increased risk of primary liver cell carcinoma.

(Burke 1986). Two errors are likely with such small numbers: either one can miss significant treatment effects or attribute efficacy when really there is none. Furthermore, "induced" seroconversion in these small groups was often compared to the seroconversion occurring in control groups with no account of confidence limits.

Table 5.3. Factors influencing spontaneous loss of HBeAg

1. Ethnic origin
2. Length of infection
3. Sex F > M
4. Sexual orientation: homosexual vs heterosexual
5. Degree of hepatic inflammatory activity: ↑ transaminases, chronic active/lobular hepatitis
6. Level of HBV replication: ↓ DNA   ↓ DNAP
7. Intercurrent infection: HIV,   ?CMV, HSV
    HAV,   delta,   ?NANB
8. Therapy: steroids
    antivirals

F, female; M, male; DNAP, DNA polymerase; HIV, human immune deficiency virus; other abbreviation in Table 5.1.

There has been much written about the possible factors influencing spontaneous loss of HBeAg (Table 5.3); however, good studies revealing statistically significant associations are few and far between. Nevertheless, the evidence that we have suggests that some of them are important and therefore, in trials of antiviral agents, large numbers of a homogeneous population should be studied and if possible some of the more important factors should be stratified for, e.g. high transaminases and now human immune deficiency virus (HIV) status.

It has been said that homosexual men respond less well to antiviral therapy. This assumption is based on a study in which the authors reviewed their antiviral experience (Novick et al. 1985). Of 19 heterosexual patients treated, 10 lost HBeAg antigen (53%) compared to only two out of 19 homosexual patients (11%). Firstly it is difficult to use the term non-response when at this time we have not demonstrated a significant response. Furthermore, the homosexual patients in this study did receive proportionally more ARA-AMP than they did alpha interferon, the heterosexual patients had higher transaminases and, although none of the homosexual men had AIDS, their HIV status was not tested because the study was carried out at a time when the first anti-HIV serology was emerging. A more recent study suggests that anti-HIV-positive homosexual men may respond less well to recombinant interferon (McDonald et al. 1987a). It would be wise to stratify for this infection in future trials.

To estimate the sample size required in a trial of antiviral therapy, one must have an accurate assessment of the spontaneous loss of HBe antigen in the population to be studied. Amongst homosexual men this was previously thought to be very low, around 2% (Viola et al. 1981), but these authors used as the denominator in their calculation the total HBsAg period of follow up, when really one should express loss of HBeAg as loss per HBeAg-positive months or years followed. A more recent study showed a spontaneous seroconversion rate of 10%, with 95% confidence limits of 2%–18% (Weller et al. 1986). If we assume a spontaneous loss of HBeAg of 10% in homosexual men per year, with various assumptions on the efficacy of the antiviral used, e.g. from a penicillin-like compound to one that is only effective in minority of patients, one can calculate the number of patients required in treatment and control groups to give a more than 95% chance of a successful experiment (Weller et al. 1986).

If we assume from the small uncontrolled studies that in the therapy of chronic hepatitis B virus infection we are not yet dealing with a penicillin-like drug then none of the uncontrolled or controlled studies published has yet included enough treated patients with appropriate controls.

Human leukocyte interferon obtained by the stimulation of buffy coat white cells with Sendai virus was soon replaced by lymphoblastoid interferon, a mixture of eight alpha interferons obtained by infection of a B-cell line (the Namalwa cell line) with Sendai virus. In more recent years, single recombinant alpha interferons have been used in clinical trials. Alpha interferons have always been an attractive proposition to those interested in antiviral therapy. They have wide-ranging antiviral effects and also effects on the immune system which may potentiate viral clearance. Recombinant gamma interferon is now available and being assessed. It is a lymphokine and, if beneficial, is likely to be useful because of its immunomodulatory effects rather than its direct antiviral effects. Of the synthetic antiviral agents, adenine arabinoside was replaced by its monophosphate ester ARA-AMP, but longer courses of this compound produced neurotoxicity in the form of peripheral neuropathy. Acyclovir has emerged as another synthetic antiviral which can inhibit HBV replication in high doses and is now being assessed in combination chemotherapy with alpha interferons.

Various methods of manipulation of the immune response have been tried but none of these has been shown to be effective. Table 5.4 shows the controlled studies of ARA-AMP that have been published. None of these trials contained sufficient numbers to make assumptions about efficacy. Unfortunately the neurotoxicity encountered with two months therapy and intravenous therapy prevented larger trials from being carried out. There have been many uncontrolled studies of alpha interferon. Lymphoblastoid interferon was shown to inhibit HBV replication as well as human leukocyte interferon (Fig. 5.7) and a thrice weekly regime inhibited HBV DNA polymerase as well as did a daily dose and this regime was better tolerated (Weller et al. 1982c; Lok et al. 1984). The randomised controlled studies of alpha

Table 5.4. Controlled studies of ARA-AMP

Reference	% loss of HBeAg	
	ARA-AMP	Control
Hoofnagle et al. 1982	20 (2/10)	20 (2/10)
- Trepo et al. 1984	55 (10/18)	26 (5/9)
Perillo et al. 1985	5 (1/18)	5 (1/17)
Weller et al. 1985	26 (4/15)	0 (0/14)

Dose 5–10 mg/day for 28 days. Ag, antigen.

interferon do provide some optimism with respect to therapy (Table 5.5). Two studies stand out perhaps as not being as successful as the others, namely those of Anderson and Lok. In Anderson et al.'s (1987) study patients were only treated for one month, whereas in the other studies two to three month courses were used. Lok et al.'s (1986) studies have been done on Chinese patients and in particular it would appear that children responded very poorly. The reasons for this are not clear. There are two European trials in progress which are extremely important (Alexander et al. 1986a, 1987; Thomas et al. 1986). The first study has excluded patients with high levels of transaminases and, of these, 9 out of 13 lost HBeAg soon after exclusion, illustrating the importance of either stratification or exclusion for this factor. In the remaining patients, although levels of HBeAg to anti-HBe seroconversion have not reached statistical significance, five patients have lost HBsAg, and this is a very unusual event in chronic HBV infection. Similarly in the second study, still in progress, several patients have lost HBsAg in the interferon group. Although neither of these studies has yet reached the numbers where one might be 95% sure of a successful experiment, the early HBsAg data suggest that the interferon-treated groups are behaving differently from those untreated patients that we have followed for many years.

Acyclovir in high dosage inhibits HBV DNA polymerase and HBV DNA, although the mechanism is obscure (Weller et al. 1983a). Controlled studies of acyclovir by intravenous infusion have shown only moderate decreases in these viral markers and no significant differences in loss of HBe antigen between treated and control patients, although again numbers of patients were small (Alexander et al. 1986b). The use of acyclovir in combination with interferon is being explored, because it may act synergistically with interferon in inhibiting HBV replication. Further information on this is required. Deoxyacyclovir is a pro-drug of acyclovir and is 100% absorbed orally. Use of this drug may facilitate longer-term therapy with acyclovir and interferon.

In the early 1980s many patients were withdrawn from steroid therapy to be entered into antiviral trials. Steroid therapy had not been shown to benefit patients with chronic hepatitis B, although it reduces considerably the morbidity and mortality of those with autoimmune chronic active hepatitis. Indeed there were studies that showed that steroids could have a deleterious effect (Lam et al. 1981). Two groups observed that, when steroids were withdrawn, spontaneous loss of HBeAg occurred shortly afterwards (Scullard et al. 1981a; Weller et al. 1982b) (Fig. 5.8). This was interpreted by some to be a passive effect, i.e. that the steroids had delayed spontaneous HBe to anti-HBe seroconversion and that when steroids were withdrawn this event was allowed to take place. Others interpret this as an active phenomenon, i.e. that the act of withdrawing steriods actively precipitated HBe to anti-HBe

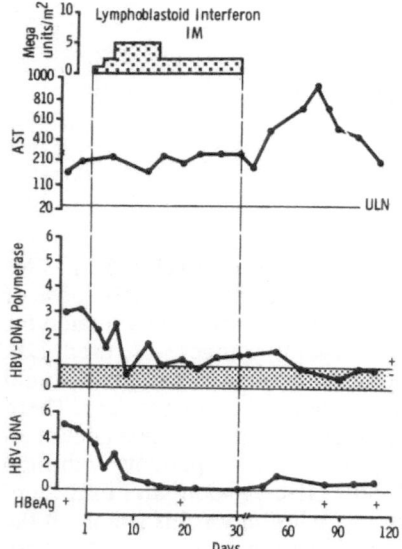

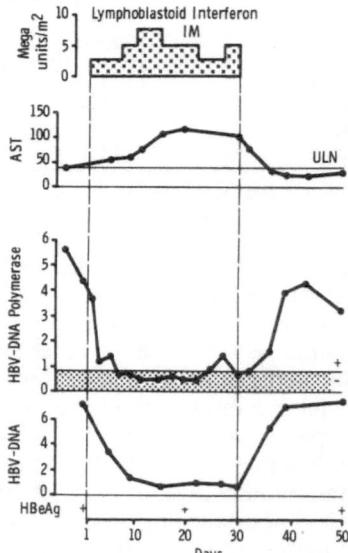

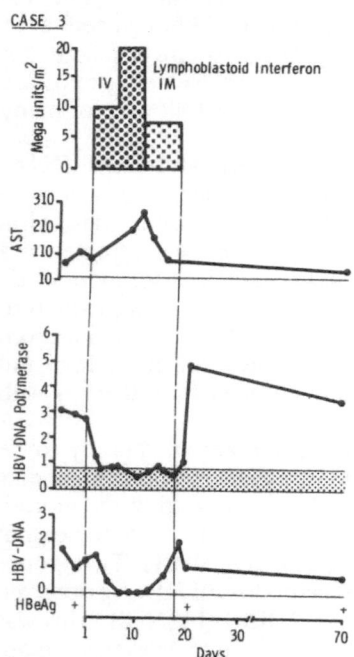

Fig. 5.7. Effect of *lymphoblastoid interferon on HBV* replication and aspartate transaminase (*AST*) (from Weller et al. 1982c). *IV,* intravenous; *IM,* intramuscular; *Ag,* antigen; *ULN,* upper limit of normal.

Table 5.5. Randomised controlled studies of alpha interferon

Reference	Type	% loss of HBeAg		Increased transaminase
		Treated	Control	
Hoofnagle 1986	Recombinant	29 (9/31)	7 (1/14)	—
Barbara et al. 1986	Lymphoblastoid	89 (8/9)	22 (2/9)	—
Anderson et al. 1987	Lymphoblastoid	15 (2/13)	0 (0/15)	
Alexander et al. 1986a, 1987	Lymphoblastoid	26 (6/23)	0 (0/23)	69 (9/13)
Thomas et al. 1986	Recombinant	19 (6/32)	0 (0/9)	—
Lok et al. 1986	Recombinant (adults)	5 (1/20)	0 (0/6)	—
Lai et al. 1987	Recombinant (children)	8 (1/12)	8 (1/12)	—

sero-conversion in those patients in whom it would not have occurred spontaneously. This has led a number of investigators to explore the use of steriod withdrawal followed by antiviral therapy. Certainly some of the studies look promising but much larger numbers are required in controlled trials (Omata and Uchiumi 1986; Perillo 1986). Futhermore, the use of steriods in HIV-infected homosexual men is not to be recommended.

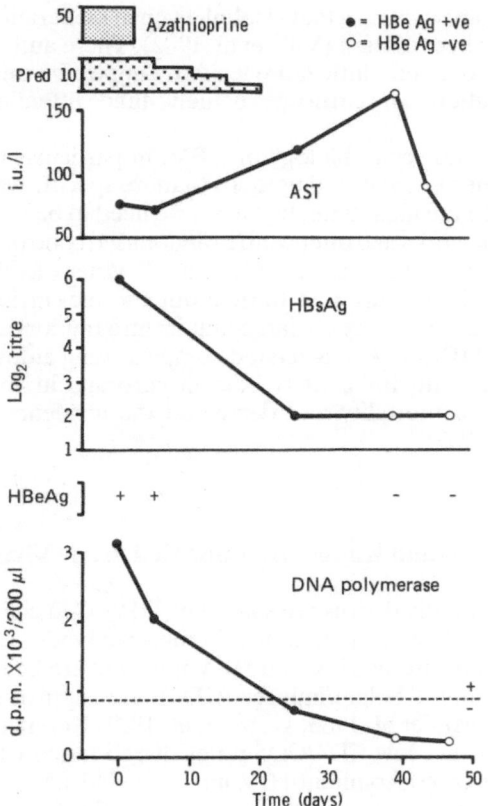

Fig. 5.8. Sequential changes in HBV markers and transaminase (*AST*) in a patient withdrawn from immunosuppressive therapy (from Weller et al. 1982b). *Pred*, prednisoline; *Ag*, antigen.

It is ironic that all these years of work in the antiviral therapy of chronic hepatitis B have led to advances not so much in therapy of hepatitis B as in the therapy of infections that we know less about. Alpha interferon has been shown to suppress hepatic inflammatory activity associated with chronic non-A, non-B virus infection (Hoofnagle et al. 1986) and chronic delta virus infection (Rizzetto et al. 1986).

In conclusion, review of the published trials of antiviral therapy in chronic hepatitis B may suggest that we have not moved very far since the mid 1970s. However, the responses in the two large randomised control trials, still underway in the UK and Europe, are providing some promising results. It is vital that these trials continue to completion to reach the sort of numbers required to show significant differences in HBeAg to anti-HBe seroconversion rates. Furthermore, trials are already underway with combination chemotherapy in the United States, with the investigators assuming that the European trials are going to show a significant benefit. If the European trials do not run to completion, these investigators will be left high and dry with respect to knowing which of their drugs is effective, if any.

In the meantime the management of the homosexual male with chronic hepatitis B is largely supportive. HBeAg-positive carriers need to be given advice and support with respect to their infectivity and regular sexual partners should be immunised if not already immune. Anti-HBe-positive carriers with normal biochemistry and histology should be reassured. There is some evidence that alcohol, even in moderation, may hasten the progression of chronic liver disease (Villa et al. 1982). These authors suggest complete abstinence; others would be a little more lenient. Supportive treatment is required for the complications of cirrhosis, namely fluid retention, encephalopathy and haemorrhage.

Advances in our knowledge of the molecular biology of HBV, in particular the characterisation of HBV DNA polymerase, will lead perhaps to more specific ways of interfering with HBV replication. In the meantime, however, we need to be pragmatic. We need to show that one agent works and multicentre randomised controlled studies are the only way of providing a quick answer. Hepatitis B virus was discovered in 1965; 23 years on we are still not sure how to treat our chronic carriers. HIV was discovered in late 1983 and four years later a large multicentre randomised controlled trial in 262 patients with AIDS or AIDS-related complex using zidovudine, a drug which had demonstrated antiviral activity both in vitro and in vivo, showed that this compound decreased mortality and decreased the incidence of opportunistic infections (Fischl et al. 1987).

Interactions between Hepatitis B Virus and Human Immune Deficiency Virus

All of the routes of transmission of hepatitis B virus are shared by HIV. Coincident infection by these viruses has therefore become common in homosexual men. The possible interactions between acute and chronic HIV and HBV infection are beginning to be examined. HBV infects both T-helper/inducer (CD4) and suppressor/cytotoxic (CD8) lymphocytes (Pasquinelli et al. 1986; Laure et al. 1987; Noonan et al. 1986). Homosexual HBV carriers have low CD4/CD8 ratios due to raised CD8 numbers. This was not accounted for by concomitant HIV infection (Novick et al. 1986). Further study of lymphocyte phenotypes and, more importantly, function is required to examine particularly the intracellular interaction between the two viruses, which may be both latently or productively infecting CD4 cells. Furthermore, chronic HBV infection could conceivably alter the natural history of HIV infection in terms of progression to clinical disease.

HIV infection may alter the natural history of HBV infection in a number of ways. Pre-existing HIV infection may alter the clinical course of acute HBV infection by affecting the incidence of fulminant hepatitis and by increasing the carrier rate following acute HBV infection. Two limited studies disagree on the prevalence of fulminant hepatitis in acute HBV infection in HIV-infected individuals (Underhill et al. 1986; Amoroso et al. 1986). In the presence of immunosuppression, the incidence of fulminant hepatitis B might be expected to be reduced because a cytotoxic T-cell response to HBV core proteins on or in hepatocytes is thought to be important in the hepatic parenchymal inflammatory response to HBV infection (Eddleston et al. 1982). Pre-existing quantitative or qualitative CD4 defects induced by HIV may decrease this response. In addition, the frequency of the chronic carrier state may be increased (Underhill et al. 1986).

Pre-existing HIV infection might be expected to impair the immune response to vaccination. In a limited study, the response rate in anti-HIV-positive homosexual men has been found to be approximately 50% compared to the usual greater than 90% (Carne et al. 1987b). In a longitudinal study of homosexual men with naturally acquired antibodies, anti-HIV-positive men were significantly more likely to have lost anti-HBs on follow up (Biggar et al. 1987).

HBV replication may be potentiated in the presence of HIV infection as shown by increased hepatocyte HBc antigen and HBe antigen with monoclonal antibody staining and increased serum HBV DNA and DNA polymerase (Perillo et al. 1986; McDonald et al. 1987b). Furthermore, the rate of spontaneous loss of HBe antigen with time may be reduced (Krogsgaard et al. 1987) and reactivation of productive viral replication may occur, although the latter may occur without concomitant HIV infection. Conversely, coincident HIV infection in the chronic HBV carrier may decrease hepatic inflammatory activity as measured by the levels of transaminases and histology (McDonald et al. 1987b; Krogsgaard et al. 1987). Finally, the possible effects of HIV coinfection and response to therapy in chronic HBV infection have already been discussed.

The Prevention of HBV Infection

The documented change in sexual behaviour in a subgroup of homosexual men in the developed world as a result of HIV infection will probably play an important role in reducing the rate of new infections (Carne et al. 1987a). Nevertheless the pool of infected carriers will remain and, without therapy of proven benefit, prophylaxis is required. Both passive and active immunisation are available for HBV infection.

It has been recommended that high-titre specific hepatitis B immunoglobulin (HBIG) is given within seven days, preferably within 48 hours of exposure to the regular sexual partners of patients with acute hepatitis B, and there is some evidence that it is effective in this situation (Redeker et al. 1975). In accidental inoculation accidents (needlestick injuries) there is considerable controversy as to whether HBIG prevents the development of infection or modifies it, allowing a transient usually subclinical course with a long incubation period followed by the production of anti-HBs (passive/active immunity) (Grady et al. 1978; Hoofnagle et al. 1979). An uncontrolled study in the UK suggested benefit with a low incidence (3%) of hepatitis (MRC and PHLS, 1980). The regular sexual partner of a homosexual man presenting with acute hepatitis B, if infected, is likely to have been so for some time before he is contact-traced. It is likely therefore that HBIG, if it has any effect, would modify rather than prevent infection. It should be given as soon as possible: a serum sample from the recipient prior to immunisation for HBsAg and anti-HBs screening is desirable but waiting for the result should never delay the administration of HBIG. There is evidence that two doses of HBIG one month apart confer greater protection than a single dose. Furthermore, since the event has identified a homosexual man who has been, and will be, at risk for HBV infection, he should also be vaccinated if he has not already been found to be immune on the pre-HBIG blood specimen.

Active Immunisation

More than 17 years ago diluted boiled serum containing HBV was given to individuals susceptible to HBV infection and it conferred immunity. Advances in knowledge with respect to the morphology of the virus, in the absence of a tissue culture system, paved the way for the separation, purification and inactivation of the 22 nm particles of HBsAg from carrier plasma. Subsequently more detailed knowledge concerning the HBV genome, its polypeptides and the amino acid sequence of HBsAg has facilitated the development of a future generation of vaccines.

The results of the first large randomised double blind control study in 1083 homosexual men in New York was reported in 1981 (Szmuness et al. 1980). Forty micrograms of a highly purified formalin-inactivated 22 nm particle preparation was given at time zero, one month and six months. Of vaccinated subjects, 95% developed anti-HBs and this persisted for the two years of follow up. The HBV attack rate was 3% in the vaccine group and 26% in the placebo group. This difference was highly significant. Many of the infections in the vaccine group were due to men being in the incubation period for HBV infection at the time of vaccination. Further studies in homosexual men have confirmed these observations (Table 5.6).

Table 5.6. Vaccine trials in homosexual men

Reference	Number	HBV events		Follow-up (months)
		Vaccine (%)	Placebo (%)	
Szmuness et al. 1980	1083	3.2	25.6	26
Francis et al. 1982	1402	8.2	17.0	18
Coutinho et al. 1983	800	4.8	23.8	21.5

Following these studies, cost–benefit analyses in the UK and the USA (Mulley et al. 1982; Adler et al. 1983) showed that considerable savings, even in respect of only the prevention of acute hepatitis, could be made to the national economies of developed world countries by offering vaccination to homosexual men. Since these studies, a vaccine prepared by the expression of the HBV genome in yeast has become available (Prinsen et al. 1987) and both this and the plasma vaccine are now approximately half of the original cost of the plasma vaccine. However, with the change in sexual behaviour, the hepatitis B attack rate is falling, but with the increased prevalence of HIV the HBV carrier rate following acute infection may be increasing. With these changing variables the original analyses are no longer relevant. Such discussion is somewhat academic anyway, since if studies were carried out on who had received the vaccine in the USA and UK, for example, it is likely that they would reveal that the bulk has gone to health-care workers. Having been given an opportunity to control infection in at-risk groups, namely homosexual men, prostitutes and intravenous drug users, it is sad that we are using the vaccine largely to protect ourselves.

I.V.D.W. is a Wellcome Trust Senior Lecturer in Infectious Diseases.

References

Adler MW, Belsey EM, McCutchan JA, Mindel A (1983) Should homosexuals be vaccinated against hepatitis B virus? Cost and benefit assessment. Br Med J 286: 1621–1624

Aldershville J, Dietrichson O, Skinhoj P, Kryger P, Mathiesen LR, Christoffersen P, Nielsen JO and the Copenhagen Hepatitis Acute Programme (1982) Chronic persistent hepatitis: serological classification and meaning of the hepatitis Be system. Hepatology 2: 243–246.

Alexander GJM, Fagan EA, Guarner P et al. (1986a) A controlled trial of 6 months thrice weekly lymphoblastoid interferon versus no therapy in chronic hepatitis B virus infection. J Hepatol 3 (suppl 2): S183–S188

Alexander GJM, Fagan EA, Hegarty JE et al. (1986b) A controlled trial of acyclovir in stable chronic HBsAg, HBeAg positive carriers. J Hepatol 3 (suppl 2): S123–S127

Alexander GJM, Brahm J, Fagan EA et al. (1987) Loss of HBsAg with interferon therapy in chronic hepatitis B virus infection. Lancet ii: 66–68

Amoroso P, Lettieri G, Giorgio A et al. (1986) Lack of correlation between fulminant form of viral hepatitis and retrovirus infection associated with the acquired immunodeficiency syndrome (AIDS) in drug addicts. Br Med J 292: 376–377

Anderson MG, Harrison TJ, Alexander G, Zuckerman AJ, Murray-Lyon IM (1987) Randomised controlled trial of lymphoblastoid interferon for chronic active hepatitis B. Gut 28: 619–622

Barbara L, Mazzella G, Baraldini M et al. (1986) Randomised controlled trial with human lymphoblastoid interferon versus no treatment in chronic hepatitis B virus infection. J Hepatol 3 (suppl 2): S235–S238

Biggar RJ, Goedert JJ, Hoofnagle J (1987) Accelerated loss of antibody to hepatitis B surface antigen among immunodeficient homosexual males infected with HIV. N Engl J Med 316: 630–631

Blumberg BS, Alter HJ, Visnich S (1965) A new antigen in leukaemia sera. JAMA 191: 541–546

Brechot C, Purcel C, Louise A, Rain B, Tiollais P (1980) Presence of integrated hepatitis B virus DNA sequences in cellular DNA of human hepatocellular carcinoma. Nature 286: 533–535

Brechot C, Hadchouel M, Scotto J et al. (1981) State of hepatitis B virus DNA in hepatocytes of patients with hepatitis B surface antigen positive and negative liver diseases. Proc Natl Acad Sci USA 78: 3906–3910

Burke CA (1986) A statistical view of clinical trials in chronic hepatitis B. J Hepatol 3 (suppl 2): S261–S267

Carne CA, Weller IVD, Johnson AM et al. (1987a) Prevalence of antibodies to human immunodeficiency virus, gonorrhoea rates and changed sexual behaviour in homosexual men in London. Lancet i: 656–658

Carne CA, Weller IVD, Waite J et al. (1987b) Impaired responsiveness of homosexual men with HIV antibodies to plasma derived hepatitis B vaccine. Br Med J 294: 866–868

Chadwick RG, Galizzi J, Heathcost J et al. (1979) Chronic persistent hepatitis: hepatitis B virus markers and histological follow-up. Gut 20: 372–377

Chakraborty PR, Ruiz-opazo N, Shouval D, Shafritz DA (1980) Identification of integrated hepatitis B virus DNA and expression of viral RNA in an HBsAg producing human hepatocellular carcinoma cell line. Nature 286: 531–533

Chalmers TC, Eckhardt RD, Reynolds WE et al. (1955) The treatment of acute infectious hepatitis. Controlled studies on the effects of diet, rest and physical reconditioning on the acute course of the disease and on the incidence of relapses and residual abnormalities. J Clin Invest 34: 1163–1235

Christenson B, Brostrom CH, Bottiger M et al. (1982) An epidemic outbreak of hepatitis A among homosexual men in Stockholm. Hepatitis A, a special hazard for the male homosexual subpopulation in Sweden. J Epidemiol 116: 599–607

Corey L, Holmes KK (1980) Sexual transmission of hepatitis A in homosexual men. Incidence and mechanism. N Engl J Med 302: 435–438

Coutinho RA, Albrecht-van lent P, Lelie N, Nagelkerke N, Kuipers H, Rijsdijk T (1983) Prevalence and incidence of hepatitis A among male homosexuals. Br Med J 287: 1743–1745

Dane DS, Cameron CH, Briggs M (1970) Virus like particles in serum of patients with Australia antigen associated hepatitis. Lancet i: 695–698

Eddleston ALWF, Mandelli M, Mieli-Vergani G, Williams R (1982) Lymphocyte cytotoxicity to autologous hepatocytes in chronic hepatitis B virus infection. Hepatology 2: 122S–127S

Edman JC, Gray P, Valenzuela P, Rall LB, Rutter WJ (1980) Integration of hepatitis B virus sequences and their expression in a human hepatoma cell. Nature 286: 535–538

Fagan EA, Williams R (1986) Serological responses to HBV infection. Gut 27: 858–867

Fischl MA, Richman DD, Grieco MH et al. (1987) The efficacy of azidothymidine (AZT) in the treatment of patients with AIDS and AIDS-related complex. A double-blind, placebo controlled trial. N Engl J Med 317: 185–191

Francis DP, Hadler SC, Thompson SE et al. (1982) The prevention of hepatitis B with vaccine. Report of the Centers for Disease Control multi-center efficacy trial among homosexual men. Ann Intern Med 97: 362–366

Grady GF, Lee VA, Prince AM et al. (1978) Hepatitis B immune globulin for accidental exposures among medical personnel: final report of a multicenter controlled trial. J Infect Dis 138: 625–638

Harville DD, Summerskill WHJ (1963) Surgery in acute hepatitis. J Am Med Assoc 184: 257–261

Heathcote J, Sherlock S (1973) Spread of acute type B hepatitis in London. Lancet i: 1468–1470

Hoofnagle JH (1980) Hepatitis B surface antigen (HBsAg) and antibody (anti-HBs). In: Bianchi L, Gerok W, Sickinger K, Stalder GA (eds) Virus and the liver. MTP Press Limited, pp 27–37

Hoofnagle J (1986) Clinical trials of the interferons. J Hepatol 3 (suppl 2): S253–S258

Hoofnagle JH, Seef LB, Bales B et al. (1979) Passive-active immunity from hepatitis B immune globulin re-analysis of a Veterans Administration Group study of needlestick hepatitis. Ann Intern Med 91: 813–818

Hoofnagle JH, Minuk GY, Dusheiko GM et al. (1982) Adenine arabinoside 5'-monophosphate treatment of chronic type B hepatitis. Hepatology 2: 784–788

Hoofnagle JH, Mullen KD, Jones B et al. (1986) Treatment of chronic non-A, non-B hepatitis with recombinant human alpha interferon – a preliminary report. N Engl J Med 315: 1575–1578

Iwarson SA (1987) Non-A, non-B hepatitis: dead ends or new horizons? Br Med J 295: 946–948

Krogsgaard K, Lindhardt BO, Nielsen JO et al. (1987) The influence of HTLV-III infection on the natural history of hepatitis B virus infection in male homosexual HBsAg carriers. Hepatology 1: 37–41

Lai CL, Lok ASF, Lin HJ, Wu PC, Yeoh EU, Yeung CY (1987) Placebo controlled trial of recombinant alpha-2 interferon in Chinese HBsAg-carrier children. Lancet ii: 877–883

Lam KC, Lai CL, Ng RP, Trepo C, Wu PC (1981) Deleterious effect of prednisolone in HBsAg positive chronic active hepatitis. N Engl J Med 304: 380–386

Laure F, Chatenoud C, Pasquinelli C et al. (1987) Frequent lymphocyte infection by hepatitis B virus in haemophiliacs. Br J Haematol 65: 181–185

Liaw YF, Chu CM, Chen TJ, Lin DY, Chang-Chien CS, Wu CS (1982) Chronic lobular hepatitis: a clinicopathological and prognostic study. Hepatology 2: 258–262

Lok ASF, Weller IVD, Karayiannis P et al. (1984) Thrice weekly lymphoblastoid interferon is effective in inhibiting hepatitis B virus replication. Liver 4: 45–49

Lok ASF, Lai CL, Wu PC (1986) Interferon therapy of chronic hepatitis B virus infection in Chinese. J Hepatol 3 (suppl 2): S209–S215

Mackay P, Lees J, Murray K (1981) The conversion of hepatitis B core antigen synthesized in E. coli into "e" antigen. J Med Virol 8: 237–243

Magnius LO, Epsmark JA (1972) New specificities in Australia antigen positive sera distinct from the Le Bouvier determinants. J Immunol 109: 1017–1021

McDonald JA, Caruso L, Karayiannis P et al. (1987a) Diminished responsiveness of male homosexual chronic hepatitis B virus carriers with HTLV-III antibodies to recombinant alpha-interferon. Hepatology 7: 719–723

McDonald JA, Harris S, Waters JA, Thomas HC (1987b) Effect of human immunodeficiency virus (HIV) infection on chronic hepatitis B viral antigen display. J Hepatol 4: 337–342

Mindel A, Tedder RS (1981) Hepatitis A in homosexuals. Br Med J 282: 1666

MRC and PHLS (Medical Research Council and Public Health Laboratory Service) (1980) The incidence of hepatitis B infection after accidental exposure and anti-HBs immunoglobulin prophylaxis. Lancet i: 6–8

Mulley AG, Silverstein MD, Dienstag JC (1982) Indications for hepatitis B vaccine, based on cost effectiveness analysis. N Engl J Med 307: 644–652

Nielsen JO, Dietrichson O, Elling P, Christoggersen P (1971) Incidence and meaning of persistence of Australia antigen in patients with acute viral hepatitis: development of chronic hepatitis. N Engl J Med 285: 1157–1160

Noonan CA, Yoffe B, Mansell PW, Melnick JL, Hollinger FB (1986) Extrachromosomal sequences of hepatitis B virus DNA in peripheral blood mononuclear cells of acquired immune deficiency syndrome patients. Proc Natl Acad Sci USA 83: 5698–5702

Novick DM, Lok SF, Thomas HC (1985) Diminished responsiveness of homosexual men to antiviral therapy for HBsAg positive chronic liver disease. J Hepatol 1: 15–27

Novick DM, Brown DJC, Lok ASF, Lloyd JC, Thomas HC (1986) Influence of sexual preference and chronic hepatitis B virus infection and T lymphocyte subsets, natural killer cell activity and suppressor cell activity. J Hepatol 3: 363–370

Omata M, Uchiumi K (1986) Combination of prednisolone withdrawal and antiviral agents (adenine, arabinoside, interferon) in chronic hepatitis B. J Hepatol 3 (suppl 2): S65–S69

Pasquinelli C, Laure F, Chatenoud L (1986) Hepatitis B virus DNA in mononuclear blood cells. A frequent event in hepatitis B surface antigen-positive and negative patients with acute and chronic liver disease. J Hepatol 3: 95–103

Perillo RP (1986) The use of corticosteroids in conjunction with antiviral therapy in chronic hepatitis B with ongoing viral replication. J Hepatol 3 (suppl 2); S57–S64

Perillo RP, Gelb L, Campbell C et al. (1979) Hepatitis BeAg, DNA polymerase activity and infection of household contacts with hepatitis B virus. Gastroenterology 76: 1319–1325

Perillo R, Regenstein F, Bodicky C et al. (1985) Comparative efficacy of adenine arabinoside 5'-monophosphate and prednisone withdrawal followed by adenine arabinoside 5'-monophosphate in the treatment of chronic active hepatitis B. Gastroenterology 88: 780–786

Perillo RP, Regenstein FG, Roodman ST (1986) Chronic hepatitis B in asymptomatic homosexual men with antibody to the human immunodeficiency virus. Ann Intern Med 105: 382–383

Pignatelli M, Waters J, Lever A, Iwarson S, Gerety R, Thomas HC (1987) Cytotoxic T-cell responses to the nucleocapsid proteins of HBV in chronic hepatitis – evidence that antibody modulation may cause protracted infection. J Hepatol 4: 11–21

Prinsen H, Goilav C, Safary A, Andre FE, Piot P (1987) Immunogenicity and tolerance of a yeast derived hepatitis B vaccine in homosexual men. Postgrad Med J 63 (suppl 2): 147–150

Redeker AG, Mosley JW, Gocke DJ, McKee AP, Pollack W (1975) Hepatitis B immune globulin as a prophylactic measure for spouses exposed to acute type B hepatitis. N Engl J Med 293: 1055–1059

Reiner NE, Judson FN, Bond WW, Francis DP, Petersen NJ (1982) Asymptomatic rectal mucosal lesions and hepatitis B surface antigen at sites of sexual contact in homosexual men with persistent hepatitis B virus infection. Ann Intern Med 96: 170–173

Repsher LH, Freebern RK (1969) Effect of exercise on recovery from infective hepatitis. N Engl J Med 281: 1393–1396

Rizzetto M, Verme G (1985) Delta hepatitis – present status. J Hepatol 1: 187–193

Rizzetto M, Rosina F, Saracco G et al. (1986) Treatment of chronic delta hepatitis with alpha-2 recombinant interferon. J Hepatol 3 (suppl 2): S229–S233

Schreeder MT, Thompson SE, Hadler SC et al. (1982) Hepatitis B in homosexual men: prevalence of infection and factors related to transmission. J Infect Dis 146: 7–15

Scott RM (1980) Experimental transmission of hepatitis B virus by semen and saliva. J Infect Dis 142: 67–71

Scullard GA, Smith CI, Merigan TC, Robinson Ws, Gregory PB (1981a) Effects of immunosuppressive therapy on viral markers in chronic active hepatitis B. Gastroenterology 81: 987–991

Scullard GH, Pollard RB, Smith JI et al. (1981b) Antiviral treatment of chronic hepatitis B virus infection. Changes in viral markers with interferon combined with adenine arabinoside. J Infect Dis 143: 772–783

Summers J (1981) Three recently described animal virus models for human hepatitis B virus. Hepatology 1: 179–183

Summers J, Mason WS (1982) Properties of the hepatitis B like viruses related to their taxonomic classification. Hepatology 2: 61S–66S

Swift WE, Gardner HT, Moore DJ, Streitfeld FH, Havens WP (1950) Clinical course of viral hepatitis and the effect of exercise during convalescence. Am J Med 8: 614–622

Szmuness W (1978) Hepatocellular carcinoma and the hepatitis B virus: evidence for a causal association. Prog Med Virol 24: 40–69

Szmuness W, Stevens CE, Harley EJ et al. (1980) Hepatitis B vaccine. Demonstration of efficacy in a controlled clinical trial in a high risk population in the United States. N Engl J Med 303: 833–841

Tedder RS, Cameron CH, Wilson-Croome R, Howell DR, Colgrove A, Barbara JAJ (1980) Contrasting patterns and frequency of antibodies to the surface, core and e antigens of hepatitis B virus in blood donors and in homosexual patients. J Med Virol 6: 323–332

Teixeira MR Jr, Weller IVD, Murray A et al. (1982) The pathology of hepatitis A in man. Liver 2: 53–60

Thomas HC, Scully LJ, McDonald JA (1986) Lymphoblastoid and recombinant alpha-A interferon therapy of chronic hepatitis B virus infection – the Royal Free Hospital experience. J Hepatol 3 (suppl 2): S193–S197

Toillais P, Charnay P, Vyas GN (1981) Biology, hepatitis B virus. Science 213: 406–411

Trepo C, Hartz O, Ouzan D (1984) Therapeutic efficacy of ARA-AMP in symptomatic HBeAg positive CAH: a randomised placebo controlled study. Hepatology 4: 1055

Underhill GS, Jeffries DJ, Forster GE, Harris JR (1986) Correlation between fulminant form of viral hepatitis and retrovirus infection associated with AIDS. Br Med J 292: 1080–1081

Villa E, Rubbiani L, Barchi T et al. (1982) Susceptibility of chronic symptomless HBsAg carriers to ethanol-induced hepatic damage. Lancet ii: 1243–1244

Viola LA, Barrison IG, Coleman JC, Paradinas FJ, Fluker JL, Murray-Lyon IM (1981) Natural history of liver disease in chronic hepatitis B surface antigen carriers – survey of 100 patients from Great Britain. Lancet ii: 1156–1159

Viola LA, Barrison IG, Coleman JC, Murray-Lyon IM (1982) The clinical course of acute type A hepatitis in chronic HBsAg carriers – a report of 3 cases. Postgrad Med J 58: 80–81

Ware AJ, Cuthbert JA, Shorey J, Gurian LE, Eigenbrodt EH, Combes B (1981) A prospective trial of steroid therapy in severe viral hepatitis. Gastroenterology 80: 219–224

Weller IVD, Bassendine MF, Craxi A et al. (1982a) Successful treatment of HBs and HBeAg positive chronic liver disease: prolonged inhibition of viral replication by highly soluble adenine arabinoside 5'-monophosphate (ARA-AMP). Gut 23: 717–723

Weller IVD, Bassendine MF, Murray A, Craxi A, Thomas HC, Sherlock S (1982b) Effects of prednisolone/azathioprine in chronic hepatitis B viral infection. Gut 23: 650–655

Weller IVD, Fowler MJF, Monjardino J, Carreno V, Thomas HC, Sherlock S (1982c) Inhibition of hepatitis B viral replication by lymphoblastoid interferon. Philos Trans R Soc Lond (Biol) 299: 128–130

Weller IVD, Carreno V, Fowler MJF et al. (1983a) Acyclovir in HBeAg-positive chronic liver disease. J Antimicrob Chemother 11: 223–231

Weller IVD, Karayiannis P, Lok ASF et al. (1983b) The significance of delta agent infection in chronic hepatitis B virus infection: a study in British carriers. Gut 24: 1016–1063

Weller IVD, Cohn D, Sierralta A et al. (1984) Clinical, biochemical, serological, histological and ultrastructural features of liver disease in drug abusers. Gut 25: 417–423

Weller IVD, Lok ASF, Mindel A et al. (1985) Randomised controlled trial of adenine arabinoside 5'-monophosphate (ARA-AMP) in chronic hepatitis B virus infection. Gut 26: 745–751

Weller IVD, Brown A, Morgan B et al. (1986) Spontaneous loss of HBeAg and the prevalence of HTLV-III/LAV infection in a cohort of homosexual hepatitis B virus carriers and the implications for antiviral therapy. J Hepatol 3 (suppl 2): S9–S16

Chapter 6

Genital Warts

J. D. Oriel

History

The first references to anogenital warts are in the literature of the ancient world, where they were described as "condylomas" or "figs" (Bafverstedt 1967). The disease seems to have been common enough to attract the attention of non-medical writers. In the later Roman Empire, promiscuous sexual behaviour, both homosexual and heterosexual, was widespread; anal warts were regarded as a clear sign of anal intercourse, and those affected were subject to ribald comments by satirical writers such as Martial (first century AD), who wrote:

In order to buy some slave boys
Labienus sold his gardens,
But now the poor man has
Only an orchard of figs.

(Epigrammata medicae philosophicae XII: 3, transl. J. D. Oriel.)

After the fall of the Roman Empire, little was written about anogenital warts until the eighteenth century. At that time syphilis, gonorrhoea and warts were common diseases, and all were attributed to the action of the same "venereal poison". Benjamin Bell (b. 1749–d. 1806) was the first to point out that gonorrhoea and syphilis were different diseases, and that warts should not be confused with the papules and condylomata lata of secondary syphilis. Despite this, an etiological muddle persisted, and in the nineteenth century warts were wrongly attributed first to gonorrhoea, then to secretions "disturbed by venery" (hence the name "venereal warts").

Towards the end of the nineteenth century, the truth slowly began to emerge. Histological resemblances between genital and common skin warts were observed, and later some skin warts were induced by inoculating filtrates of genital wart tissue into non-genital epithelia (Oriel 1971a). It was then believed that genital and skin warts were closely related. For many years the latter had been suspected as being due to a virus, and support for this view came in the late 1940s, when virus particles were demonstrated by electron microscopy in extracts of common skin warts. Twenty years later, similar particles were seen in extracts of both anal and genital warts; it was assumed that all varieties of wart were caused by the same virus, but research during the last decade has shown that this opinion is incorrect.

The concept that genital warts in heterosexuals are a sexually transmitted disease was at first only slowly accepted, but has now been confirmed (Oriel 1971a). Although anal warts are very common in homosexual men they have received surprisingly little attention. For many years the association with anoreceptive intercourse that the ancient Romans had observed was not mentioned – indeed, at the end of the nineteenth century anal warts were described as almost exclusively a disease of women. In the twentieth century, perhaps because in many countries homosexual acts were formerly illegal, references to the pathogenesis of anal warts in men were scanty; eventually, a tacit belief that they were related to anal intercourse received confirmation from epidemiological studies (Oriel 1971b), but even today the virology and natural history of the disease are in many respects uncertain.

Virology

Human papillomaviruses (HPV) are a genus of the family Papovaviridae. The virion is 50–55 nm in diameter; there is no capsule. The genome is a double-stranded molecule of molecular weight 5×10^6; the capsid shows icosahedral symmetry and is composed of 72 protein subunits (McCance 1986).

Replication of papillomaviruses takes place in a differentiating squamous epithelium. Viral DNA can be detected in all cell layers, but assembly of complete virions occurs only in the keratinising epithelium towards the surface of a wart. No papillomavirus has yet been propagated in vitro, and for this reason lack of viral antigen has made it impossible to differentiate HPV types by classical serological methods. It is now, however, possible to divide the genus into types according to the homology of DNA molecules extracted from the lesions (McCance 1986). At present, at least 40 HPV types are recognised, of which HPV6, 11, 16, 18, 31, 33 and 35 have been found in anogenital material. In men, DNA sequences of HPV6 and 11 are found predominantly in anogenital condylomata acuminata, and only rarely in dysplastic or malignant lesions. In contrast, sequences of HPV16 and 18 are present in the majority of penile and anal carcinomas and their precursors, and are uncommon in benign penile and anal warts.

Pathology

All warts share basic histological features. The dermal papillae are elongated. The basal cell layer is intact, and the stratum spinosum hyperplastic (acanthosis). The granular cell layer is thickened, and the stratum corneum may be hyperplastic. In the stratum granulosum there are some large vacuolated cells whose nuclei contain basophilic inclusions: these are shown by electron microscopy to consist of aggregates of virus particles. Similar aggregates are present in the stratum corneum.

In condyloma acuminatum, which is the usual variety of anogenital wart, all these features are present. Acanthosis is particularly well developed, constituting the bulk of the tumour, but there is little hyperkeratosis, unless the warts have aged or been unsuccessfully treated. In sessile and common warts, which often occur on the shaft of the penis, hyperkeratosis is much more marked.

Non-condylomatous wart virus infection of the uterine cervix is very common (Meisels et al. 1982), but can be identified clinically only by colposcopy. After treatment with acetic acid, the cervix shows areas of "acetowhite" epithelium with or without a granular surface, which have been called flat condylomas. Cervical cytology shows the presence of koilocytes, cells which have large clear cytoplasmic vacuoles and irregular and hyperchromatic nuclei; multiple nuclei are present in some of the cells. Koilocytes are regarded as pathognomonic of wart virus infection. Dyskeratosis is seen in many of the specimens, but is not a specific indicator of the condition. It has been suggested that subclinical lesions resembling flat condylomas may also occur on the vulva and penis (Levine et al. 1984).

Medley (1984) examined Papanicolaou-stained anal smears taken from the vicinity of the dentate line in a group of homosexual men, many of whom had a history or current evidence of perianal warts, and stated that he found cytological evidence of wart virus infection in 44% of the men. Frazer et al. (1986) have suggested that the anatomy and histology of the dentate line are analogous to those of the uterine cervix, with a "transformation zone" at the junction of squamous and columnar epithelium. They defined cytological criteria for the diagnosis of HPV infection in rectal smears, which included koilocytosis, with or without multinucleation, and dyskeratosis.

Immunology

Little is known about humoral or cell-mediated immunity (CMI) to HPV infections. Circulating antibodies to a group antigen have been detected in women with genital warts (Baird 1983), but lack of adequate amounts of viral antigen makes it impossible to look for type-specific antibodies. Studies with immunosuppressed patients have shown the importance of immune responses in determining the outcome of HPV-induced disease. Thus, women who are immunosuppressed after renal transplantation are very liable to condylomata acuminata and women with lymphomas are subject to vulval warts that are not only intractable but may progress to intra-epithelial neoplasia (Schneider et al. 1983).

In homosexual men, CMI may be defective because of HIV infection, but it is also defective in some men without HIV infection, particularly those who practise frequent anal intercourse (Detels et al. 1983; Frazer et al. 1986); the reasons for this are unknown. Immune deficiency may contribute towards the development of extensive refractory anal condylomata and to the appearance of oral "hairy" leucoplakia, which may be an unusual manifestation of HPV infection. It may also underlie the progression of some benign HPV-induced lesions to ano-rectal dysplasia (Frazer et al. 1986).

Epidemiology

Incidence

There is little reliable information on the incidence of HPV infection in homosexual men. In England, condylomata acuminata are reportable from sexually transmitted disease (STD) clinics but the number of infections in heterosexual, bisexual and homosexual men are not reported separately. Many years ago, Oriel (1971b) reported that, of 500 unselected homosexual men attending an STD clinic in London, 1.4% had penile warts and 10% anal warts. Carr and William (1977) reported from New York that 22% of a group of 682 homosexual men stated that they either had anal warts or had had them in the past. The average age of onset of anal warts in both the above studies was 23 years.

Infectivity

Anal warts are associated with anoreceptive intercourse, yet in homosexual men they are many times more common than penile warts (Oriel 1971b; Carr and William 1977). The reasons for this are uncertain. Perhaps the warmth and moisture of the perianal environment preferentially favours the growth of the tumours, but it is also possible that the immunosuppression that occurs in anoreceptive individuals (whether or not they have HIV infection) may enhance the growth of warts in the ano-rectum. There are no data available on the infectivity of either penile or anal warts in homosexual men.

Associated Infections

Sexually transmitted infections are common in homosexual men, including those with anogenital warts, and careful screening for these infections is essential. As a minimum, urethral, rectal and pharyngeal culture for *Neisseria gonorrhoeae* and serological tests for syphilis should be performed before the warts are treated (see Chap. 2).

Clinical Features

Penis and Scrotum

Genital warts are polymorphic. *Condylomata acuminata* are soft fleshy vascular tumours; they may be single, but are more often multiple and can coalesce into large masses. Although any part of the penis and scrotum may be affected, these condylomas often begin on areas that are subject to trauma during intercourse, such as the frenum, corona and inner lining of the prepuce. *Sessile warts* are slightly raised lesions, 1–2 mm in diameter, which affect dry areas of epithelium, particularly the shaft of the penis; they can occur alone or associated with condylomata acuminata elsewhere. *Common warts* are sometimes seen on the shaft of the penis, often accompanying similar warts on non-genital skin.

The urethra is involved in at least 10% of men with HPV infection of the penis; it may be the only site affected. Nearly 80% of tumours are in the distal 3 cm of the urethra, but proximal spread occurs occasionally (de Benedictis et al. 1977). Urethral warts are of condyloma acuminatum type, and because the stratum corneum is very thin they are bright red and bleed easily. Patients may complain of urethral bleeding, dysuria, or urethral discharge, but the disease is often symptomless.

Subclinical HPV infection of the penis is common. Multiple "microwarts" may be seen if magnification is used; further, colposcopy of the penis after the application of 5% acetic acid often shows areas of acetowhite epithelium invisible to the naked eye, occurring either alone or on the periphery of clinically apparent lesions. The terminal urethra may also be the site of inconspicuous warts (Levine et al. 1984). The existence of "silent" HPV infection has obvious epidemiological and therapeutic implications.

Anus and Anal Canal

Perianal warts are usually condylomata acuminata. They are multicentric, and may reach a substantial size. Internal condylomas affect over 50% of men with perianal warts (Schlappner and Shaffer 1978); most of these are in the anal canal, but they can occur above the pectinate line. Cytological studies have indicated that occult ano-rectal HPV infection may be common in homosexual men. Law et al. (1986) found that 54 of 100 homosexuals or bisexuals had cytological or histological evidence of HPV infection of the anal canal or rectum, although fewer than 20% had visible condylomata acuminata.

Oral Cavity

Condylomata acuminata have been recorded on the lips, tongue and palate (Oriel 1986). Some patients have concurrent genital or anal warts, and others give a history of oral sex with partners with anogenital HPV infection.

A new form of oral leucoplakia has been described affecting homosexual men, many of whom are immunosuppressed because of HIV infection (Greenspan et al. 1984). The lesions affect the sides of the tongue and have a corrugated or "hairy" surface. They have the histological features of warts, and HPV antigen has been demonstrated in the majority. Epstein-Barr virus is also present in most of the lesions. The pathogenesis, course and outcome of the disease are unknown, but it does appear to be, at least in part, a manifestation of HPV infection.

Differential Diagnosis

Penile warts must be distinguished from other papular lesions of the area. *Fordyce's spots* (Fordyce's syndrome) are ectopic sebaceous glands, and appear as multiple tiny yellow papules, often affecting the inner lining of the prepuce. *Pearly penile papules* (hirsutes papillaris penis) are hypertrophic papillae with a normal epidermal covering. They appear as parallel rows of filiform lesions on the corona. *Sebaceous cysts* are common on the shaft of the penis and scrotum.

Condylomata lata of secondary syphilis are rounded lesions that may appear on the penis; other signs of early syphilis are present, and both dark-field examination for *Treponema pallidum* and syphilis serology are positive (see Chap. 7). The domed lesions of *molluscum contagiosum* have a characteristically umbilicated appearance. In tropical countries donovanosis (granuloma inguinale) may affect the penis; the demonstration of Donovan bodies is diagnostic.

The distinction of penile warts from premalignant and malignant tumours is of the utmost importance, but may be difficult because dysplastic lesions may accompany, or develop from, benign viral condylomas. *Bowenoid papulosis* is a premalignant disease that presents as multiple reddish-brown papules, some of which are verrucoid and may resemble warts. *Carcinoma of the penis*, like HPV infection, commonly involves the glans, coronal sulcus and prepuce and, particularly if it is of the "cauliflower" form, may give diagnostic problems. Early biopsy, repeated if necessary, is essential for any atypical penile lesion.

The differential diagnosis of anal warts is from fibroepithelial polyps, anal lesions of syphilis (particularly condylomata lata), and from premalignant and malignant neoplasia.

Complications

Giant Condyloma

Giant condyloma, the eponymous Buschke–Löwenstein tumour, affects the penis and, more rarely, the anus. It begins as a small warty lesion, usually in the coronal sulcus; it extends relentlessly, with the destruction of large areas of the penis and the formation of multiple fistulae. Clinically it appears to be a florid carcinoma, but it does not metastasise and histologically it is benign, consisting only of condyloma acuminatum tissue. The nature of the disease is uncertain. It may be essentially a benign viral tumour, a concept supported by the identification of DNA sequences of HPV6 in some specimens (Gissmann et al. 1982); its aggressive behaviour would then be due to unknown additional factors. On the other hand, it has been argued that giant condyloma is probably a well-differentiated carcinoma from the outset, the development of squamous cell carcinoma that has been recorded in some cases representing simply a loss of differentiation (Tessler and Applebaum 1982).

Premalignant and Malignant Disease

The association between papillomavirus infection, particularly with HPV16, and cervical and vulval neoplasia is now well documented (Oriel 1986), and evidence is accumulating of a similar association in the male genital tract. Penile intra-epithelial neoplasia presents with lesions that may be single or multiple, papular or erosive, and pigmented or non-pigmented. Premalignant lesions include erythroplasia of Queyrat, Bowen's disease and Bowenoid papulosis; histologically they all have the features of carcinoma in situ. Clinically it may be difficult to differentiate some of these lesions from condyloma acuminatum (Chapel and Rahbari 1980), and indeed there is an antecedent history of genital warts in many patients (Wade et al. 1979). DNA sequences of HPV16 have been reported in 80% of cases of Bowenoid papulosis of the penis (Ikenberg et al. 1983).

Cancer of the penis is rare in Europe and the United States, but in South America it is much commoner. McCance et al. (1986) have studied a series of penile carcinomas from South America, and have reported HPV16 DNA sequences, integrated into the host cell chromosomes, in approximately half the specimens. It may be concluded from these studies that papillomaviruses, particularly HPV16, may have a causal or co-causal role in both premalignant and malignant tumours of the penis. The sexual orientation of the patients studied has not been reported, so it is not possible at present to indicate whether or not these diseases pose a particular threat to homosexual men.

It has recently been suggested that there may be an association between ano-rectal dysplasia and HPV in male homosexuals. Several epidemiological studies have indicated that anal (but not rectal) cancer is many times commoner in homosexual than in heterosexual men (Austin 1982; Daling et al. 1982; Peters and Mack 1983); the results would be consistent with a hypothesis that anal coitus may be a risk factor for anal cancer.

Both intra-epithelial neoplasia and invasive carcinoma arising in anal warts have been described (Oriel and Whimster 1971; Ejeckam et al. 1983). Frazer et al. (1986) have reported an association between cytological features of dysplasia and of HPV infection in some rectal smears from homosexual men, and that in these men dysplasia may persist for at least a year. Dysplasia was associated with a history of anal warts, frequent receptive anal intercourse and the presence of serum antibody to HIV. The results of these studies suggest the possibility that, as in the female genital tract, HPV infection may predispose to anal and rectal dysplasia. However, the HPV types involved have not yet been defined, the natural history of ano-rectal dysplasia and its possible progression to carcinoma in situ and invasive cancer, and the role of other agents and immunological mechanisms in these events, are alike unknown. For these reasons judgement on the involvement of HPV in ano-rectal neoplasia in homosexual men must at present be reserved.

Treatment

Before treatment of anogenital warts is started, it is important to make sure that the diagnosis is correct, to define the extent of the disease and to exclude concomitant infections.

Accurate Diagnosis

If the tumours are typical condylomata acuminata, plane or common warts, most physicians start treatment without biopsy. Although perhaps undesirable in principle, in practice this approach works well, but it should again be emphasised that any atypical lesion should be examined histologically before treatment is begun.

Extent of Disease

The penis should be carefully examined in a strong light, preferably with magnification. Previous application of 5% acetic acid helps to show acetowhite areas suggestive of HPV infection. The urinary meatus should be closely inspected. After the perianal area has been examined, proctoscopy is performed and the mucosa of the anal canal and rectum searched for condylomas; rectal cytology is not at present a routine investigation.

Finally, the scrotum, groins and pubic areas are inspected. Many men who complain of penile warts have anal lesions, and vice versa, so the above scheme of examination should be followed in every case.

Concomitant Infections

These should be excluded before treatment. As a minimum, urethral, rectal and pharyngeal culture for *N. gonorrhoeae*, and serological tests for syphilis, should be performed. Men who are immunocompromised, whether because of HIV infection or for some other reason, may respond poorly to treatment.

There is no uniformly effective treatment for anogenital warts, and a therapeutic regimen should be decided for each patient, to be varied if it is not proving effective.

Medical Treatment

Podophyllin resin is extracted from the rhizomes of *Podophyllum peltatum* and *P. emodi* and contains several cytotoxic agents of variable potency. A 10% or 25% (w/v) solution in ethanol or benzoin tincture is carefully applied to the warts and allowed to dry. Initially, it is washed off after four hours, but this interval can be lengthened to 24 hours with successive applications. Treatment is continued once or twice a week until the lesions have regressed.

Podophyllin has several disadvantages, and in some countries it is not used at all. Its composition is not standardised, and response to therapy is variable and often poor. It is particularly ineffective against perianal warts, and should not be applied to condylomata in the anal canal, for its actions there cannot be controlled. It is irritating to normal skin, which means that it must be applied by trained personnel and not by the patient. Further, podophyllin is potentially toxic. Small quantities are well tolerated, but liberal application to large condylomatous masses has been followed by absorption and systemic effects that include dizziness, vomiting, peripheral neuritis and even coma and death (Montaldi et al. 1974). A further possibility is that in some circumstances podophyllin may be oncogenic (Karol et al. 1980); certainly, the histology of warts that have been treated with podophyllin and later removed is very difficult to interpret. For these reasons podophyllin, if it is to be used, must be used with caution. Not more than 0.5 ml of a 25% solution should be applied at any one treatment session, and a course of treatment limited to three to four weeks. If cure has not been achieved by this time, another form of treatment should be substituted.

Many other cytotoxic agents have been tried for the treatment of anogenital warts, but none has become established. Topical antiviral agents such as idoxuridine and acyclovir are ineffective. Interferons make cells resistant to virus infection and also have effects on cell growth and maturation and on the immune system. They show activity against HPV-induced lesions, although the mechanisms are not known. The injection of α-2b interferon into genital warts induces significant regression (Eron et al. 1987); multiple injections are required, and there are side effects, which include fever, myalgia and leucopenia. Systemic α- and β-interferon has been used for the treatment of extensive condylomata acuminata with some effect, but major side effects have been encountered (M. Campion, personal communication). The place of these drugs in the treatment of anogenital warts is not yet decided.

Surgical Treatment

Surgery is essential for:

1. Extensive penile or anal warts
2. Intrameatal condylomas whose extent cannot be clearly seen
3. Condylomas in the anal canal
4. Warts which have not responded to medical treatment

The following procedures are available.

Cautery

In areas such as the shaft of the penis where local anaesthesia is practicable, small warts can be removed by electrocautery and curettage. Inadequate treatment will probably be followed by recurrence, but too vigorous cautery can lead to scarring. Electrodesiccation by high frequency sparking with a "Hyfrecator" is precise and effective. A dental syringe with a fine needle is ideal for injecting the local anaesthetic. Neither electrocautery nor electrodesiccation is suitable for extensive warts.

Cryotherapy

Liquid nitrogen, a carbon dioxide snow stick or a cryosurgical probe operated by nitrous oxide may all be used, but the last of these is the most convenient. Each lesion must be frozen until its whole thickness has been penetrated; the skin folds should be stretched so that adjacent normal skin is spared (Bunney 1982). In many cases, retreatment will be required every week or two until all the lesions have disappeared. Again, cryotherapy is not the best technique for large or extensive condylomas.

Diathermy

This is a good technique for extensive warts, and can also be used for intrameatal and intra-anal lesions. A general anaesthetic is necessary. Diathermy is destructive, and scarring after healing may be a problem; anal stenosis has developed in some men after perianal warts have been treated in this way.

Laser Therapy

A carbon dioxide laser is used. Small warts can be treated under local anaesthesia, but a general anaesthetic is necessary for extensive disease. Laser causes destruction of the warts by evaporation of water from the cells. The depth and breadth of the area treated can be accurately gauged, and healing occurs from a healthy tissue base, so is usually rapid. Laser therapy is suitable for both penile and anal condylomas.

Excision

When there is extensive or persistent condylomatous involvement of the prepuce, circumcision may be indicated. Thomson and Grace (1978) have devised a scissor excision technique for perianal and intra-anal warts; scarring is radial, so anal stenosis does not occur. This procedure has been adapted by McMillan and Scott (1987) for the treatment of perianal lesions under local anaesthesia.

References

Austin DF (1982) Etiological clues from descriptive epidemiology: squamous carcinoma of the rectum or anus. Nat Cancer Inst Monogr 62: 89–90

Bafverstedt B (1967) Condylomata acuminata – past and present. Acta Dermatol Venereol (Stockh) 47: 376–381

Baird PJ (1983) Serological evidence for the association of papillomavirus and cervical neoplasia. Lancet ii: 17–18

Bunney MJ (1982) Viral warts. Their biology and treatment. Oxford University Press, New York, Toronto, pp 48–55

Carr G, William DC (1977) Anal warts in a population of gay men in New York City. Sex Transm Dis 4: 56–57

Chapel TA, Rahbari H (1980) Genital Bowenoid papulosis – squamous cell carcinoma in situ. Sex Transm Dis 7: 139–141

Daling JR, Weiss NS, Klopfenstein LL et al. (1982) Correlates of homosexual behaviour and the incidence of anal cancer. JAMA 247: 1988–1990

de Benedictis TJ, Marmar JL, Praiss DE (1977) Intraurethral condylomata acuminata: management and a review of the literature. J Urol 118: 767–769

Detels R, Fahey JL, Schwartz K et al. (1983) Relation between sexual practices and T-cell subsets in homosexually active men. Lancet i: 609–611

Ejeckam GC, Idikio HA, Nayak V, Gardiner JP (1983) Malignant transformation in an anal condyloma acuminatum. Can J Surg 26: 170–173

Eron LJ, Judson F, Tucker S et al. (1987) Interferon therapy for condylomata acuminata. N Engl J Med 315: 1059–1064

Frazer IH, Crapper RM, Medley G et al. (1986) Association between anorectal dysplasia, human papillomavirus and human immunodeficiency virus in homosexual men. Lancet ii: 657–660

Gissmann L, de Villiers EM, zur Hausen H (1982) Analysis of human genital warts (condylomata acuminata) and other genital tumours for human papillomavirus type 6 DNA. Int J Cancer 29: 143–146

Greenspan D, Greenspan JS, Conant M et al. (1984) Oral "hairy" leucoplakia in male homosexuals: evidence of association with both papillomavirus and a herpes-group virus. Lancet ii: 831–834

Ikenberg H, Gissmann L, Gross G et al. (1983) Human papillomavirus type 16 related DNA in genital Bowen's disease and in bowenoid papulosis. Int J Cancer 32: 563–565

Karol MD, Conner CS, Watabe AS et al. (1980) Podophyllum: suspected teratogenicity from topical application. Clin Toxicol 16: 283–286

Law C, Medley G, Cossartt YE et al. (1986) Diagnosis of occult anal human papillomavirus (HPV) infection in homosexual males based on anal cytology, histology and DNA probe analysis. Second World Congress of Sexually Transmitted Diseases, Paris, 25–28 June: Abstract 20–07

Levine RU, Crum CP, Herman E, Silvers D, Ferenczy A, Richart RM (1984) Cervical papillomavirus infection and intraepithelial neoplasia: a study of male sexual partners. Obstet Gynecol 64: 16–20

McCance DJ (1986) Genital papillomavirus infections: virology. In: Oriel JD, Harris JRW (eds) Recent advances in sexually transmitted diseases no. 3. Churchill Livingstone, Edinburgh, pp 109–125

McCance DJ, Kalache A, Ashdown K et al. (1986) Human papillomavirus types 16 and 18 in carcinomas of the penis in Brazil. Int J Cancer 37: 55–60

McMillan A, Scott GR (1987) Outpatient treatment of perianal warts by scissor excision. Genitourin Med 63: 114–115

Medley G (1984) Anal smear test to diagnose occult anorectal infection with human papillomavirus in men. Br J Vener Dis 60: 205

Meisels A, Morin C, Casas-Cordero M (1982) Human papillomavirus infection of the cervix. Int J Gynecol Pathol 1: 75–94

Montaldi DH, Giambrone JP, Courey NG et al. (1974) Podophyllin poisoning associated with the treatment of condylomata acuminata: a case report. Am J Obstet Gynecol 119: 1130–1131

Oriel JD (1971a) Natural history of genital warts. Br J Vener Dis 47: 1–13

Oriel JD (1971b) Anal warts and anal coitus. Br J Vener Dis 47: 373–376

Oriel JD (1986) Genital papillomavirus infections: clinical manifestations. In: Oriel JD, Harris JRW (eds) Recent advances in sexually transmitted diseases, no. 3. Churchill Livingstone Edinburgh, pp 127–145

Oriel JD, Whimster I (1971) Carcinoma in situ associated with virus-containing anal warts. Br J Dermatol 84: 71–73

Peters RK, Mack TM (1983) Patterns of anal carcinoma by gender and marital status in Los Angeles County. Br J Cancer 48: 629–636

Schlappner OLA, Shaffer EA (1978) Anorectal condylomata acuminata: a missed part of the condyloma spectrum. Can Med Assoc J 118: 172–173

Schneider V, Kay S, Lee HM (1983) Immunosuppression as a high-risk factor in the development of acuminatum condyloma and squamous neoplasia of the cervix. Acta Cytol (Baltimore) 27: 220–224

Tessler AN, Applebaum SM (1982) The Buschke-Löwenstein tumour. Urology 20: 36–39

Thomson JFS, Grace RH (1978) Treatment of perianal and anal condylomata acuminata. A new operative technique. JR Soc Med 71: 181–185

Wade TR, Kopf AW, Ackerman AB (1979) Bowenoid papulosis of the genitalia. Arch Dermatol 115: 306–308

Chapter 7

Syphilis

J. S. Bingham

Syphilis is a fascinating disease that was common until a few decades ago. Untreated, it may run a prolonged course and its manifestations are many and varied. It is different from most other sexually transmitted diseases in that it is systemic from the outset. In the past, the disease has been more prevalent that it is today and it has had its influence on the course of events by its effects on many persons who have played important roles in the history of the Western world.

Epidemiology

Syphilis maintained a high incidence at the end of the nineteenth and at the beginning of the twentieth centuries. Considerable mortality was associated with this. In 1900, the Registrar General recorded 1639 adult deaths and 1200 infant deaths due to venereal diseases in England and Wales. These were probably not very accurate figures and Sir William Osler put the truer figure at around 60 000; many of these would have been due to syphilis. In the United Kingdom, the highest incidence, since figures have been kept, was recorded in 1946 (27 791) (Communicable Disease Surveillance Centre, 1986, personal communication). With the advent of penicillin, the incidence thereafter declined dramatically until the late 1950s when the incidence curve flattened out. In the United States, cases of primary and secondary syphilis dropped from over 35 000 cases per year to a low of approximately 6500 cases in 1956.

After the liberalisation of laws relating to homosexuality in the 1960s, some homosexual men developed promiscuous sexual life-styles. While not true for the whole world, in the large conurbations of Europe, Britain and America, early infectious syphilis began to be seen more in this group. Syphilis cases in males in the United Kingdom, remained steady at between 2000 and 3000 cases per year, until the 1970s, when the number rose to a peak of 3765 in 1978 and then declined to 2594 in 1984

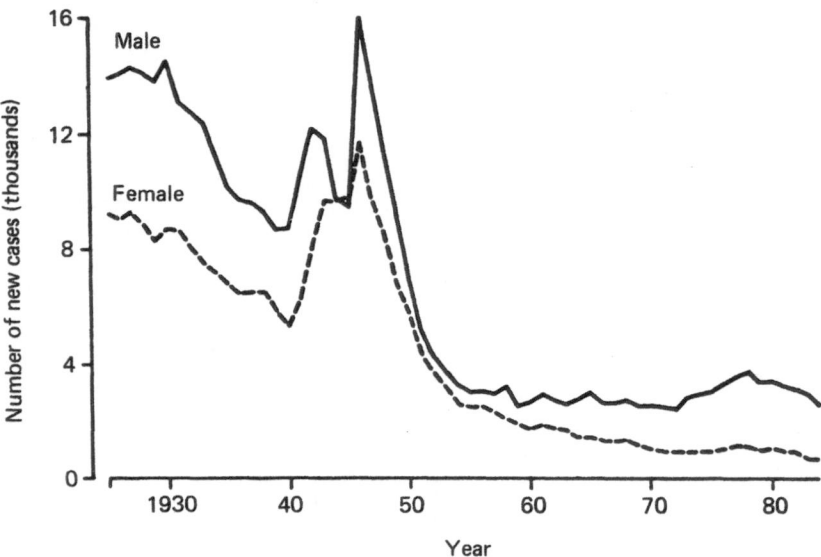

Fig. 7.1. Incidence of new cases of syphilis in the United Kingdom (Communicable Disease Surveillance Centre 1986).

(Fig. 7.1). Much of this rise in incidence has been attributed to homosexual transmission. In the four Thames (London) Health Authorities, 60.6% of cases of syphilis reported in 1984 were in men (Department of Health and Social Security 1985). The British Co-operative Clinical Group (1980) reported on 1363 cases of syphilis studied in 1977 and showed that, in Wales, Northern Ireland and Scotland, the percentage of cases in homosexual men was 14.0, 22.2 and 32.4 respectively. In England alone (excluding London), the figure was 52.2% but, in five central London Genito-Urinary Medicine Departments, 76.9% of cases were in homosexual men. At the Middlesex Hospital, in the Bloomsbury Health District, in 1985, 80% of the cases of early infectious syphilis occurred in homosexual or bisexual men. Indeed, if only primary and secondary syphilis were considered, between 1980 and 1984, 96% of cases occurred in this group. In a 20 year retrospective study at The Middlesex Hospital, changes in the epidemiology of syphilis have been highlighted in the clinic population. In the late 1960s the "average" male patient with primary syphilis was equally likely to be British or non-British and homosexual or heterosexual; 50% of these men gave a history of a previous sexually transmitted disease (STD). In the early 1980s the "average" patient was homosexual, British, had suffered from multiple previous STDs and had many sexual partners (A. Mindel, S. J. Tovey and P. Williams 1987, personal communication).

Nearly half of the reported cases of early infectious syphilis in the United States of America occurred in homosexual or bisexual men (Judson 1977). A striking change in the sex ratio (males : females) among primary and secondary syphilis cases was noted, increasing from 1.5 : 1 in 1967 to 3.2 : 1 in 1980 (Centers for Disease Control 1984). The percentage of white infected men who reported at least one male partner increased from 38% in 1969 to 70% in 1979. In Denmark in 1979, at least 48%, and possibly as much as 71% of early infectious syphilis was homosexually acquired

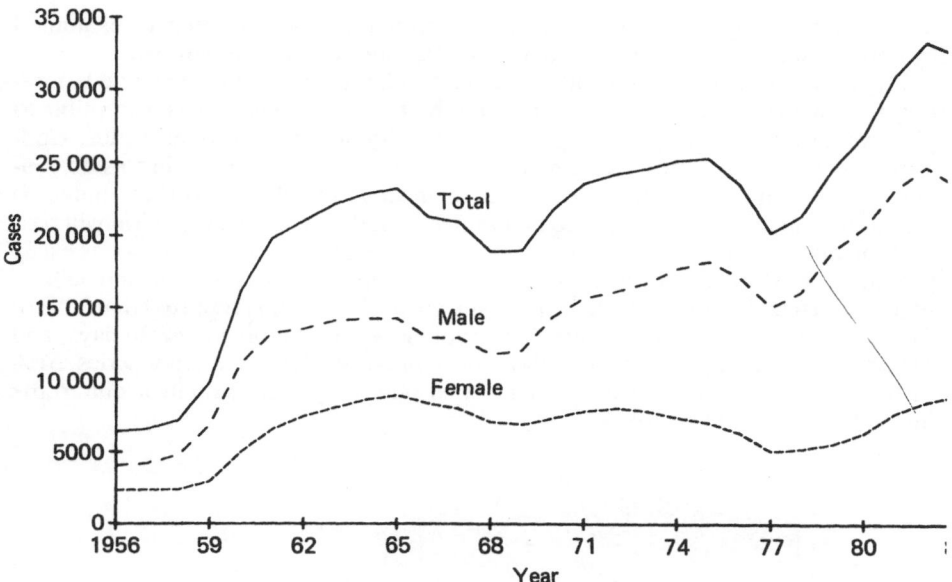

Fig. 7.2. Incidence of cases of syphilis in the United States of America (Centers for Disease Control 1984).

(Peterson and Pederson 1983). Thus, at least in large cities in the industrialised world, early infectious syphilis is seen predominantly in homosexual men. It is not surprising therefore that, with the advent of human immunodeficiency virus infection and the alteration in sexual behaviour and decline in promiscuity with which it has been associated, the incidence of syphilis has fallen. At The Middlesex Hospital, between 1981 and 1985, there has been a 34% decline in the incidence of early infectious syphilis, and in the United Kingdom in males with primary or secondary syphilis there has been a 48% decline in cases, from 1457 in 1980 to 699 in 1985 (Communicable Disease Surveillance Centre, personal communication). The overall incidence in the United States (Fig. 7.2) is down from its 1982 peak of 14.6 cases to 12.4 cases per 100 000 of the population in 1984, mainly as a result of fewer cases in homosexual men (Judson 1986). A declining incidence of syphilis among homosexual men in Stockholm has also been observed from almost 18 in 1982 to about 3 per 100 000 in 1985 (von Krogh et al. 1986).

The Causative Organism

Syphilis is caused by the spirochaete *Treponema pallidum* (Fig. 7.3). The genus *Treponema* contains other forms that are also pathogenic, such as the organisms that cause yaws, bejel and pinta. At present, for routine purposes, these organisms are indistinguishable morphologically and serologically from *T. pallidum*. The organism itself has a thin spiral form, from 6–14 µm in length. It is not easily stained and cannot be visualised by light microscopy. Instead, material from a lesion may be examined by dark-ground microscopy for evidence of these motile spiral organisms.

Treponema has never been cultured outside a human or animal host and laboratory treponemes are usually maintained in rabbit testes. *T. pallidum* is susceptible to physical and chemical agents, e.g. heat, cold, drying, and soap and water. Consequently syphilis is almost exclusively sexually transmitted; direct skin-to-skin contact, skin to mucous membrane contact or mucous membrane to skin contact is required from an infected lesion to an uninfected individual. The organism will pass through abraded skin and intact mucous membranes, but the disease is not as easily passed on as gonorrhoea. Early syphilis is transmitted to between 10% and 50% of sexual contacts. For example, Schroeter and co-workers (1971) reported that 30% of 57 sexual contacts exposed to infectious syphilis within the preceding 30 days, and observed untreated for another 90 days, developed syphilis. In another series, 51% of the contacts of persons with primary or secondary syphilis actually acquired the infection (Schober et al. 1983).

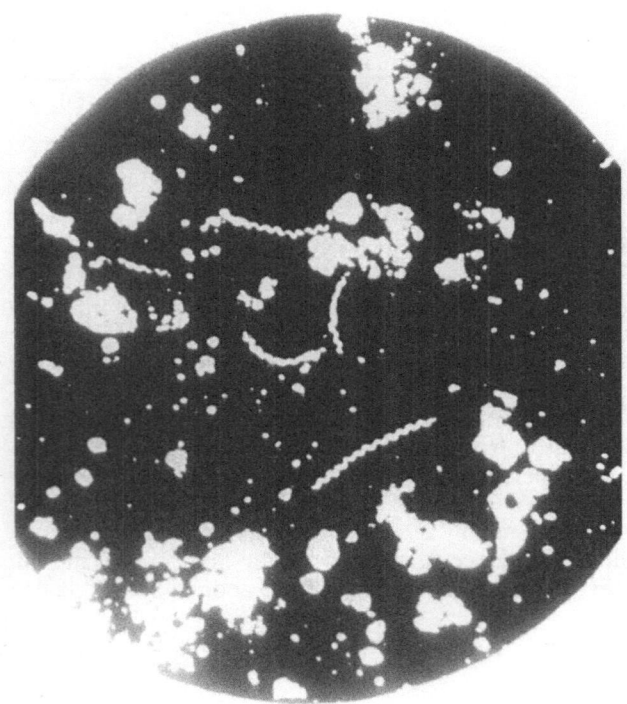

Fig. 7.3. Dark-ground photomicrograph of *Treponema pallidum*.

Clinical Aspects

The clinical manifestations of acquired infectious syphilis have not changed appreciably over the years; it is an extremely variable disease and classic descriptions in older textbooks cannot be bettered (Stokes et al. 1945).

The first manifestation is the appearance of a primary chancre, denoting the advent of the *primary stage* of the disease. Six to eight weeks after the beginning of the primary stage, the first symptoms and signs of the *secondary stage* may appear, although they may appear sooner or, occasionally, very much later. It is important to note that both the primary and secondary stages may be so inconspicuous and so transient that they may not be noticed by the patient. About 25% of patients diagnosed in the secondary stage give no history of a primary lesion and show no residual signs of it (King and Nicol 1975). When the signs of the secondary stage disappear, the patient has entered the *early latent stage*. At this stage there are no symptoms nor signs of the disease and in some untreated cases, or where treatment has been inadequate, a *recurrent phase* may ensue when the manifestations of the secondary stage reappear. This does not usually occur after the end of the second year of infection when the early infectious phase of syphilis ends and the patients pass into the *late latent stage*. Subsequently some patients may develop gummatous syphilis (benign tertiary syphilis) or cardiovascular or neurosyphilis.

The late forms of syphilis are not often seen now and, whereas cardiovascular syphilis and the parenchymatous forms of neurosyphilis were seen by most medical undergraduates 20 years ago, this is no longer the case. Affection of the central nervous system in the secondary stage of the disease occasionally presents, as does also meningovascular neurosyphilis but these are rarities. So in young homosexual men the spectrum of the disease encompasses early infectious syphilis, latent syphilis and certain forms of neurosyphilis, and only these areas will be discussed (Table 7.1).

Table 7.1. Classification of syphilis

A. Congenital syphilis	
B. Acquired syphilis	
Early infectious	
Primary	– 9–90 days after exposure
Secondary	– 6 weeks–6 months (4–8 weeks after primary lesion) after exposure
Early latent	– from end of secondary stage until 2 years after the onset of the primary lesion
Late non-infectious	
Late latent	– from 2 years after the onset of the primary lesion
Gummatous syphilis	– 3–12 years after infection
Neurosyphilis	– 3–20 years after infection
Cardiovascularsyphilis	– 10–20 years after infection

Primary Syphilis

The primary chancre, the lesion of primary syphilis, appears between nine and 90 days after contact with an infected person, the most common incubation period being around three weeks. Chancres develop at the site of treponemal inoculation and, in homosexual men, are most commonly found in the genital area or in the ano-rectal region. Other extragenital chancres can occur. The classic chancre (Fig. 7.4) begins as a dull red macule, which becomes papular and subsequently the surface erodes giving rise to an ulcer that is rounded or oval in shape and has a clearly defined border. The chancre becomes indurated, hard and usually painless (unless secondarily infected) and, if the surface is cleaned, considerable serous exudation occurs. This

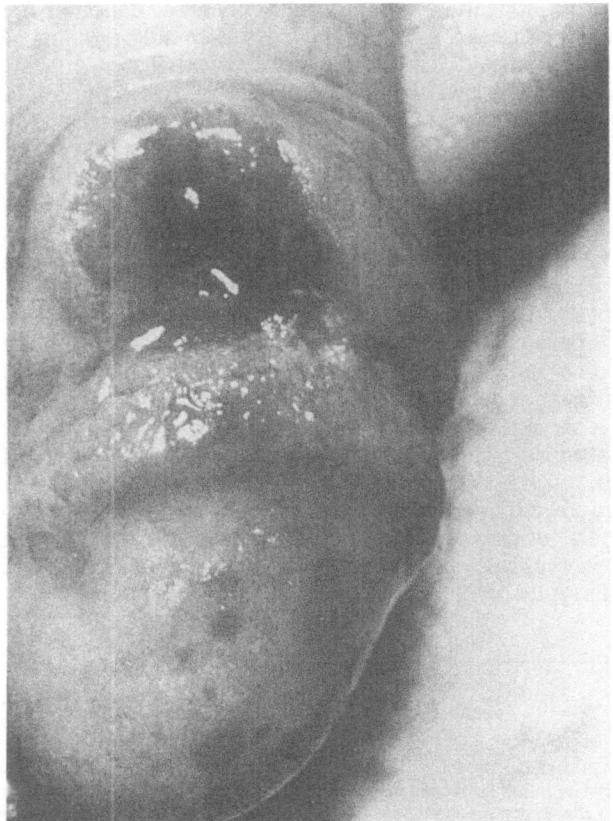

Fig. 7.4. A classic penile chancre.

contains numerous treponemes and is highly infectious. Oedema of the surrounding tissues is common. Chancres are usually single but may be multiple. Untreated, a chancre will persist for two to eight weeks and then disappear spontaneously.

Regional lymphadenopathy normally accompanies a chancre and usually develops one week after the appearance of the initial lesion. The commonest sites for chancres, the penis and the anus, drain to the superficial inguinal glands and the adenopathy at this site consists of painless, discrete, firm nodes with a rubbery consistency on at least one side.

Penile chancres are most commonly found in the coronal sulcus but can be found on the shaft, the prepuce, the glans, the frenum and in the urethral meatus. Anal chancres (Fig. 7.5) are usually visible at the anal margin, especially when the anus is relaxed, and rectal chancres (King and Nicol 1975) are rare, but may produce symptoms such as rectal pain, tenesmus, diarrhoea or constipation and occasionally bleeding. There may be femoral lymph node enlargement, or no lymphadenopathy with rectal lesions, and dark-ground examination looking for *T. pallidum* may be difficult to assess as there are many non-pathogenic spirochaetes present in the bowel (Quinn et al. 1981). Whether or not proctitis occurs as a result of infection with *T. pallidum* is uncertain.

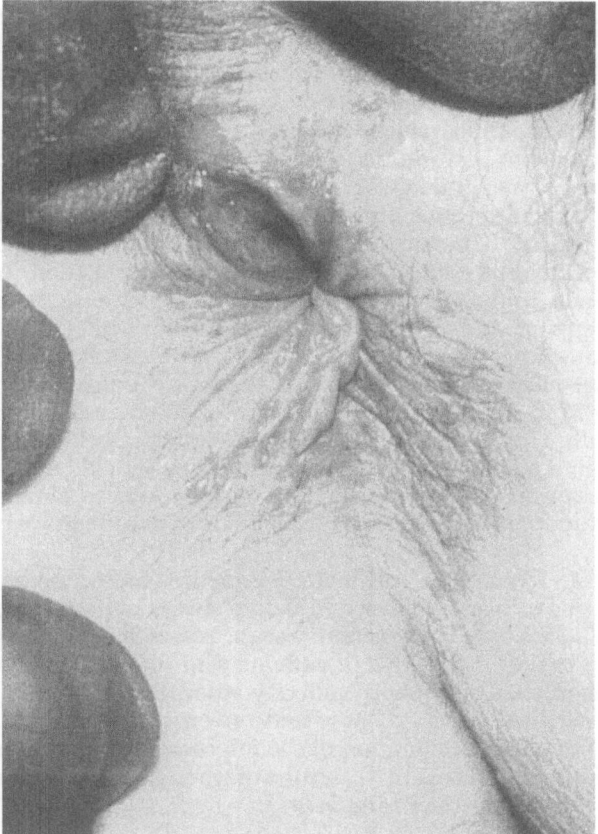

Fig. 7.5. An anal chancre.

Other extragenital chancres may appear on the finger, where it may assume a wide array of forms (Knox and Rudolph 1984), in the mouth (lips, buccal mucosa, tonsils and tongue) or at any site where the treponeme may be inoculated as a result of trauma, such as a bite.

Diagnosis

The diagnosis is suggested by the history and the clinical appearances. If extragenital sites are involved, the diagnosis can be missed. Confirmation of the diagnosis is by laboratory investigation: dark-ground examination of material from a lesion and serological tests for syphilis.

It is important to prepare properly for collection of a specimen suitable for dark-ground microscopy. Diagnostic sensitivity is greatest in specimens taken from active, weeping chancres, mucous patches and condylomata lata (in secondary syphilis). It is important to clean the surface with a saline-soaked gauge swab. This has the effect of producing a considerable serous exudate from the base of the lesion which, if due to syphilis, will be teeming with *T. pallidum*. If a lot of exudate is present this may be dabbed directly on to a glass slide or, more usually, if less exudate exists, material obtained with a scarifier can be placed on to a drop of normal saline on a slide and covered with a coverslip. The dark-ground condenser cuts out the central light and the treponemes are illuminated by peripheral light reflected upwards to the objective. It has always been traditional to perform three examinations but it is wise to realise the importance of proper preparation – a good specimen may reveal the organism whereas three poorly taken specimens may not. The specificity of the test depends very much on the experience of the microscopist, who will note *T. pallidum* with its corkscrew-type rotation about the long axis and characteristic angulation. A positive test is absolute evidence for the disease, but a negative test, because there may only be a few organisms present, does not exclude the diagnosis of syphilis. False-positive examinations may occur when *Fusobacterium* spp., which are present in the multiple shallow, tender ulcers of anaerobic balanoposthitis. Confusion can arise in the sampling of lesions in the buccal cavity because of commensal spiral organisms already in the mouth. However, if the lesion is morphologically like that of syphilis and is not on the gum margins, dark-ground microscopy should be performed. In cases where the lesion is unsuitable for taking a dark-ground specimen, aspiration of a regional lymph node and dark-ground examination of its contents may be undertaken.

Serological tests for syphilis (STS) provide only indirect evidence of infection but can be of assistance in diagnosing primary syphilis, and should always be performed on any patient with genital ulceration. A venereal disease research laboratory (VDRL) slide test is positive in only 50%–70% of patients with primary syphilis (Wende et al. 1971) and the fluorescent treponemal antibody-absorption (FTA-Abs) test in 70%–90% (Duncan et al. 1974). This is because seroconversion does not usually occur until about three weeks after acquisition of the infection, but a chancre may appear before that. In dark-ground negative cases the diagnosis is more secure if repeat quantitative VDRL test demonstrate a rise in titre.

Differential Diagnosis

Syphilis is a potentially serious disease. All genital ulceration must be investigated to exclude that diagnosis. The commonest cause of genital ulceration in the Western world is genital herpes. In classical cases it differs from syphilis by presenting with multiple vesicular eruptions that subsequently become small erosions, which are superficial, non-indurated and often tender. In primary cases there may be discrete lymph gland enlargement, which is often tender. This is the major infection to be differentiated from primary syphilis, although the two can coexist both on the penis and at the anus. It is important not to forget traumatic ulcers, especially where anal intercourse is taking place. Chancroid and granuloma inguinale are rarely seen in homosexual men but it is important to establish in which country sexual activity has taken place, as a tropical venue might raise the index of suspicion for these diseases. Lymphogranuloma venerum may produce a proctitis but external ulceration is generally transitory and not usually seen by the doctor. (For differential diagnosis of genital ulceration (see Table 3.2, p. 48).

The main differential diagnosis, apart from herpes, at the anus, is that of an anal fissure (Drusin et al. 1976) fistulae and haemorrhoids have also been misdiagnosed.

Confusion with an anal fissure is the commonest mistake and has led to surgical excision. Anal fissures are extremely painful, may bleed, and it is often almost impossible to pass a proctoscope whereas an anal chancre is relatively painless unless secondarily infected, and rarely bleeds. With a chancre there will usually be inguinal lymphadenopathy and there may be evidence of secondary syphilis.

In the mouth, the differential diagnosis includes herpes simplex infection, the mucous patches of secondary syphilis and aphthous stomatitis and Behçet's syndrome.

Secondary Syphilis

During the incubation period and the primary stage of syphilis, treponemes multiply and disseminate in the blood throughout the body. When sufficient multiplication has taken place, secondary syphilis becomes manifest (four to eight weeks after the primary lesion). A wide variety of symptoms and signs may present.

Constitutional symptoms and signs can occur, the most notable of which are malaise, fever, headache, a sore throat, hoarseness, anorexia and generalised aching. These manifestations, if present, are usually mild and transitory but can be severe. Generalised lymphadenopathy is common, with discrete, firm, rubbery, non-tender easily palpable nodes (50%–60% of cases).

The skin lesions of secondary syphilis are, perhaps, the best-known manifestations and are present in 75%–80% of cases. The rash (syphilide) is usually non-itchy, symmetrical and is pleomorphic; lesions may be macular, papular, papulosquamous or pustular, but the majority are maculopapular. The eruptions are usually indolent and may persist for weeks or months if untreated. Often the palms and soles (Fig. 7.6) are involved and this can be helpful in making a diagnosis, since few other conditions produce this appearance. In areas where papular lesions are moist, especially the perianal region, some become elevated (condylomata lata) and these are the most infectious lesions of syphilis (Fig. 7.7). Temporary patchy hair loss on the scalp, and elsewhere, may occur in association with papular lesions. On the other hand, in some cases, rashes may be inconspicuous and transient; patients in the latent stage of syphilis can often give no history of having had secondary syphilis.

Mucous patches (flattened, erosive lesions of mucous membranes) occur in the secondary stage of syphilis, most commonly in the mouth (Fig. 7.8), but may also affect the glans penis in uncircumcised men. These lesions are painless, may be transitory but can last for two to three weeks and are highly infectious.

A plethora of other manifestations may occur in secondary syphilis. Although the central nervous system is rarely involved in early syphilis, acute syphilitic meningitis usually associated with cranial nerve palsies may occur; nerve deafness has been reported (Willcox and Goodwin 1971), and anterior uveitis. Hepatomegaly may be present in up to 10% of patients (Knox and Rudolph 1984). Hepatitis may develop (Feher et al. 1975) and a disproportionately high alkaline phosphatase level may be suggestive of a syphilitic aetiology (Keisler et al. 1982). Proctitis (McMillan and Lee 1981; Akdamar et al. 1977), a rectal mass (Quinn et al. 1982), erosive gastritis (Sachar et al. 1974) and pulmonary nodules (Geer et al. 1985) have also been described. Arthritis, bursitis, periostitis, osteitis and myositis have also been reported, as well as the nephrotic syndrome. Descriptions of these conditions are given in the standard texts (Stokes et al. 1945; King and Nicol 1975; Knox and Rudolph 1984).

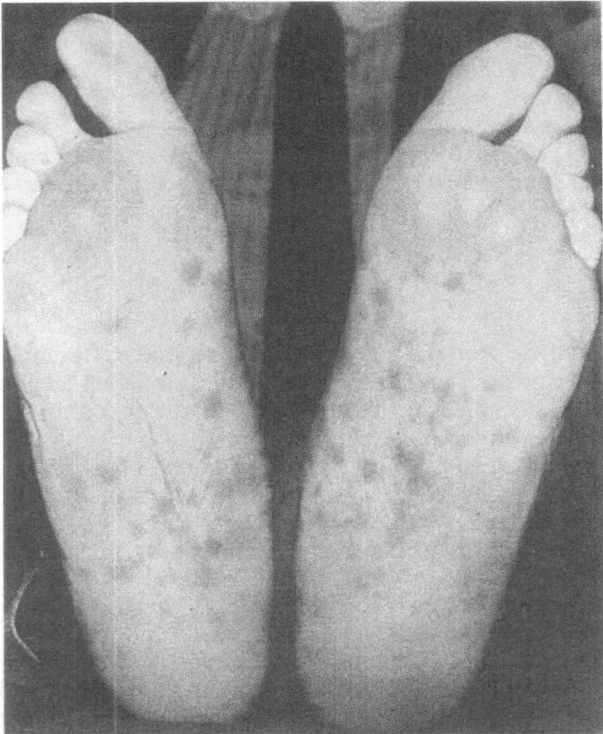

Fig. 7.6. Papular lesions of secondary syphilis on the soles of the feet.

Diagnosis and Differential Diagnosis

Because secondary syphilis can present in such a multiplicity of ways and can mimic a variety of skin disorders, STS should be performed if the diagnosis is suspected. When characteristic skin and mucous membranes lesions are present and the STS are positive, a presumptive diagnosis of syphilis can be made and the patient should be treated forthwith. The STS are always positive in secondary syphilis, so negative tests exclude the diagnosis. Dark-ground examination should be performed on mucous lesions and condylomata lata.

Cutaneous eruptions can be confused with pityriasis rosea, lichen planus, psoriasis, drug eruptions, rubella, measles, infectious mononucleosis, seborrheoic dermatitis and exanthemata. A skin biopsy should be carried out in cases of doubt. Buccal lesions have to be distinguished from aphthous ulceration, Behçet's syndrome, erythema multiforme, pemphigus and lichen planus, as well as herpes simplex infection. The constitutional symptoms of secondary syphilis could be confused with the symptoms of the AIDS-related complex (ARC) (Smego et al. 1984).

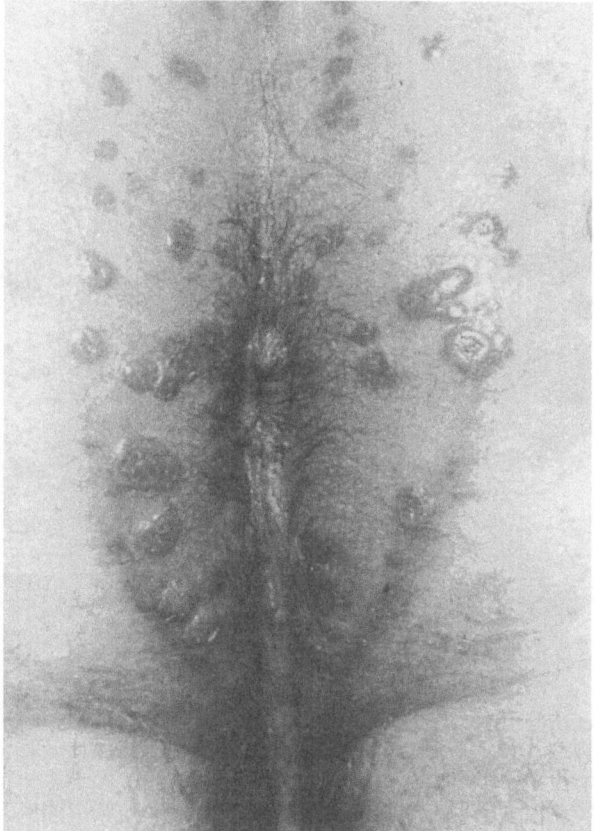

Fig. 7.7. Perianal condylomata lata.

Latent Syphilis

Where a patient is found to have positive STS, but does not give a history of having had symptoms or signs suggestive of syphilis, it is possible that he may be in the latent stage. If within the preceding two years the patient has been known to have had negative STS, or has had a significant rise in reagin titre, then he has early latent syphilis. If this is not the case then the disease cannot be staged without examination of the cerebrospinal fluid (CSF) and cardiovascular investigations. An abnormality may indicate a diagnosis of late syphilis rather than latent syphilis. This is rare in young homosexual men. However, early infectious syphilis often passes unnoticed in homosexual men and presentation in the latent stage is not uncommon.

Diagnosis

The diagnosis of latent syphilis is usually made in homosexual men when they present for a routine check-up or as a result of contact tracing action. Where it is not possible

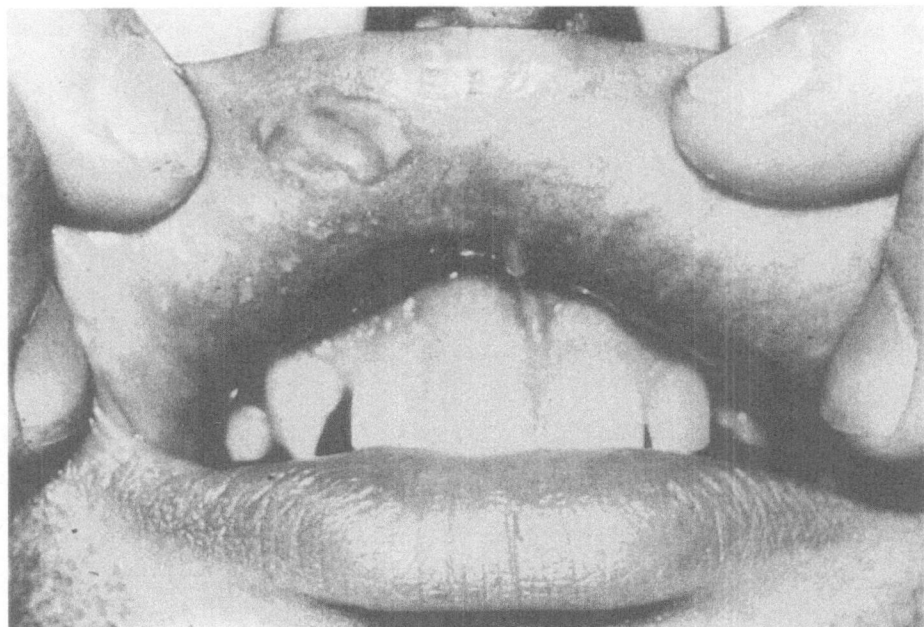

Fig. 7.8. A mucous patch on the lip.

to ascertain the duration of the infection further investigations should be undertaken. The CSF should be examined for evidence of neurosyphilis and the estimations that may be performed, in order of increasing specificity, include: total protein, cell count, presence of plasma cells and/or reactive lymphocytes, IgG as a percentage of total protein IgG index, oligoclonal IgG, TPHA index, oligoclonal IgM and oligoclonal immunofixation of treponemal IgM bands (Johnson and Thompson 1986). Cardiovascular investigations, particularly if aortic disease is suspected, might include a chest X-ray (postero-anterior), fluoroscopic screening of the aorta and echocardiography.

Neurosyphilis

It is wise not to forget that neurosyphilis can occasionally occur in young men. The possible clinical presentations are diverse, may present difficulty with diagnosis and will not be described in detail in this chapter. Evidence of asymptomatic neurosyphilis may be detected on examination of the CSF (see above). Acute meningitis can occur within the first two years of infection, but mostly during the first six months and usually associated with the secondary stage of syphilis. The meningitis is principally a basilar type and cranial nerve palsies, especially affecting the third, fourth and sixth cranial nerves are found. Headache is common but the signs and symptoms are not usually as marked as in an acute bacterial meningitis. Convulsions may occur.

Meningovascular neurosyphilis may also present within the first two to five years of infection. The signs and symptoms will vary depending on the site involved, spinal cord or cerebrum, and the degree of involvement of the blood vessels and cranial nerves. The middle cerebral artery and its branches are most commonly affected. Cranial nerve and pupillary abnormalities occur and some patients present with convulsions.

Natural History

The natural history of untreated syphilis has been studied. A large prospective study of patients with early infectious syphilis, diagnosed on clinical grounds alone, was conducted in Oslo from 1890 to 1910. The patients were not given treatment on the grounds that, in 1890, it seemed that the treatment available (primarily mercurials) was more toxic than the disease. Almost 2000 cases were enrolled and nearly 50 years later follow-up information was obtained in almost 80% of the study group (Gjestland 1955). Of untreated patients, 23.6% experienced a secondary relapse, showing that waxing and waning was a normal part of the course of untreated syphilis. Of these secondary relapses 90% occurred within the first year and 94% within two years. Gummatous syphilis was the commonest late form of the disease, affecting 14.4% of males and 16.7 of females. Some 14.9% of males and 8.0% of females developed cardiovascular syphilis. Neurosyphilis developed in 9.4% of males and 5.0% of females (Table 7.2). The probability of dying directly as a result of untreated syphilis in the Oslo study was 17.1% in males and 8.0% in females, after 40 years of infection.

Table 7.2. Studies on the natural history of syphilis

	Frequency		
	Oslo study (N = 887)		Tuskagee study (at 20 years follow-up) (N = 159)
Secondary syphilis relapse	23.6		—
	Males (N = 303)	Females (N = 584)	
Gummatous syphilis	14.4	16.7	4.0
Recognised cardiovascular syphilis	14.9	8.0	6.0
Recognised neurosyphilis	9.4	5.0	4.0

N, no. of patients.

In 1932, a second natural history study was begun by the US Public Health Service: the Tuskegee Study (Rockwell et al. 1964). At the 20 year follow-up point, 4.0% of cases had developed gummatous syphilis, 6.0% cardiovascular syphilis and 4.0% neurosyphilis. The findings showed an increased frequency of hypertension and deaths, not due to specific lesions of late syphilis.

A third study was conducted by Paul Rosahn (1947). This autopsy showed that at least 60% of untreated patients did not develop late recognisable anatomical complications and about 80% did not die primarily because of syphilis.

Because the majority of homosexual men in metropolitan areas do present for regular check-up examinations and investigations, most cases of syphilis are treated in the early infectious stage. Effective contact tracing identifies other cases in the early stages. So the late forms of the disease are only rarely seen in homosexual men and mortality associated with this group is virtually non-existent.

Serological Tests for Syphilis

There are two main types of STS available: a test measuring an IgG antibody in the serum (reagin), and specific treponemal tests. Of the former, the VDRL, the RPR (rapid plasma reagin) card tests and the ART (automated reagin test) are the ones in common use. Of the latter, the TPHA (*Treponema pallidum* haemagglutination assay) and the FTA-Abs are used. Because laboratory errors occasionally happen and because biological false-positive reactions (BFPR) and a prozone phenomenon (Spangler et al. 1964) may occur with reagin tests, both a reagin and a treponemal test should always be performed (Table 7.3). The TPHA is the usual specific test to be carried out and is semi-automated. The FTA-Abs need only be used if there is difficulty in interpreting the results. The FTA-Abs is the first of the STS to become positive, usually around three to four weeks after infection, followed by the reagin tests and the TPHA.

Table 7.3. Serological tests for syphilis

Non-specific tests (measure reagin)
Venereal disease research laboratory test (VDRL)
Rapid plasma reagin test (RPR)
Automated reagin test (ART)

Specific (treponemal) tests
Treponema pallidum haemagglutination assay (TPHA)
Fluorescent treponemal antibody-absorbed test (FTA-Abs)
Treponema pallidum immobilisation test (TPI) – rarely used
   now

An acute BFPR may last a few weeks or months and may be associated with infections when they produce fever, and with malaria and with infectious mononucleosis. Chronic reactions lasting for six months or more may be found in association with connective tissue disorders such as systemic lupus erythematosus and rheumatoid arthritis as well as other auto-immune disorders, leprosy and in intravenous drug abusers. This type of reaction is more common in females than in males.

Reagin tests should always be quantitatively titred before treatment as a base line for follow-up tests. The titre usually falls after treatment. Specific treponemal tests will generally remain positive for the rest of the patient's life, despite treatment, providing evidence of past infection when reagin tests have reverted to negativity. STS do not differentiate between other treponematoses and syphilis.

Treatment

Treponema pallidum does not appear to have developed resistance to antibiotics. Provided that the patient is not allergic to penicillin that remains the preferred drug for the treatment of all stages of syphilis. For early infectious syphilis, aqueous procaine-penicillin G (600 000 units daily by intramuscular injection for ten days) is the treatment of choice (Dunlop 1985). This regimen has been used for many years in the United Kingdom with success. It does have the disadvantage that the patient has to attend for injection daily and this is unacceptable to some patients. Because of this, in some countries, especially in the United States of America, the recommended treatment for early infectious syphilis is benzathine penicillin G (2.4 million units, 1.2 million units in each buttock, by intramuscular injection). Some believe that this should be repeated one week later (Felman 1983). This has also produced good results over the years. Intramuscular benzathine penicillin forms a repository that breaks down slowly over 15–20 days, releasing penicillin G (benzyl penicillin) into the blood. Serum levels are usually maintained above the treponemicidal level (0.5–1.5 µg/ml).

Spiral forms have been known to persist in lymphatic and other tissues after penicillin treatment (Collart et al. 1964). However, in 1976 Tramont described two patients, one with neurosyphilis and the other with secondary syphilis, who had been treated with recommended doses of benzathine penicillin. Despite this therapy in both cases *T. pallidum* was isolated from the CSF. Others have shown that benzathine penicillin cannot be relied upon to produce adequate CSF levels (Ovčinnikov et al. 1973; McDermott 1958; Smith et al. 1956; Short et al. 1966; Mohr et al. 1976). The finding that benzathine penicillin does not reach the CSF does not indicate that it has failed to cure the disease. Also, finding some treponemes in the CSF does not mean that the patient has active or progressive disease. Given that treponemes disseminate widely throughout the body in the secondary stage of syphilis, should benzathine penicillin be used at all rather than aqueous penicillin? Long experience with benzathine penicillin for the treatment of early infectious syphilis has proved very satisfactory, but it should not be used to treat neurosyphilis. Procaine penicillin (1.8 million units by intramuscular injection daily for 21 days) with probenecid (500 mg four times daily orally) has been recommended and does produce adequate treponemicidal levels in the CSF. This is a suitable treatment regimen for the late forms of syphilis (Dunlop 1985).

At The Middlesex Hospital, benzathine penicillin is only used in patients who are thought to be unlikely to attend for a daily course of injections, and also in those who for personal reasons are unable to complete a course at that particular time; in this latter group, retreatment with aqueous procaine penicillin G is advocated at a later, more convenient time.

In those who are penicillin-allergic, treatment with a tetracycline is indicated. Doxycycline is the tetracycline of choice (Dunlop 1985) and can be administered daily in a single 300 mg dose (with milk after eating) for 15 days. In complicated or late syphilis, a 21 day course is appropriate. Where the patient cannot tolerate penicillin or doxycycline, erythromycin (500 mg (with milk after eating) four times a day for 21 days) may be used. It crosses natural barriers poorly and therefore hardly enters the CSF and the eye.

While patients virtually always respond to the treatment described, retreatment may be necessary in the following circumstances:

Where the original treatment was known to be suboptimal or details are unknown
Where clinical signs and symptoms recur (rare and usually due to reinfection)
Where there is a sustained fourfold increase in titre for a reagin test, at follow-up
Where an initially high-titre reagin test fails to decrease four-fold within one year

Patients with early infectious syphilis should be followed up for two years after treatment. If they are well, and a satisfactory serological response has been obtained, then they may be discharged. Cases of late syphilis should be followed up for life.

Patients should be told not to have sex until treatment is completed. Sexual contacts should be sought, whenever possible, and examination and investigations for syphilis performed. Some patients do not wish to be treated epidemiologically (empirical routine treatment, before or without investigation) and prefer instead to be followed until the end of the incubation period and only be treated if infection has actually occurred. Epidemiological treatment is as for early infectious syphilis. Schroeter and co-workers (1971) showed that aqueous procaine penicillin G (4.8 million units with or without 1 g probenecid), used to treat gonorrhoea, was 100% effective in aborting incubating syphilis in 117 patients. It seems reasonable to assume that statim doses of ampicillin and amoxycillin as well as seven day courses of tetracycline should do the same (Judson 1986). So, if a patient is being treated as a syphilis contact then, provided adequate gonococcal therapy is administered (not spectinomycin), if syphilis is incubating it should be aborted and further treponemicidal therapy should not be necessary (Judson 1986).

Finally, patients being treated for early syphilis should be warned about the Jarisch–Herxheimer reaction, which may occur within 3–12 hours after the start of treatment, whatever form it takes. Symptoms, especially a secondary rash, may become worse, and there may be fever, rigors, headache and malaise. It may last for up to six hours and does not recur later in treatment. Putkonen and co-workers (1966) found the reaction to occur in 55% of cases with seronegative primary syphilis and in about 95% with seropositive primary or secondary syphilis. This can be a most frightening episode for the patient if he is not forewarned.

Conclusion

The incidence of syphilis is declining in the large cities of the industrialised world. Homosexual men remain the main group of patients to acquire the infection in this setting. The late forms of syphilis are rarely seen now and treatment to the point of presumptive cure is usual and relapse is rare. Whether or not treponemes are eliminated by conventional treatment is not established. Promiscuous homosexual men should be encouraged to present themselves for screening on a three monthly basis when, apart from STS, they can be screened for other sexually transmitted infections. These visits also provide an opportunity for the provision of information about the various diseases. Given that spiral organisms are known to persist after treatment of syphilis (Collart et al. 1964), it is interesting to postulate that the disease might recrudesce in the presence of immune deficiency associated with the human immunodeficiency virus (HIV). Many patients with HIV infection have previously had syphilis and there have been recent case reports of neurosyphilis developing in some who have had conventional therapy for early infectious syphilis (Berry et al. 1987; Johns et al. 1987). Neurosyphilis should probably be added to the growing list of infectious complications of the acquired immune deficiency syndrome and may be one of the earliest complications to appear.

References

Akdamar K, Martin RT, Schinose H (1977) Syphilitic proctitis. Digestive Dis 22: 701–704

Berry CD, Hooton TM, Collier AC et al. (1987) Neurologic relapse after benzathine penicillin therapy for secondary syphilis in a patient with HIV infection. N Engl J Med 316: 1587–1589

British Co-operative Clinical Group (1980) Homosexuality and Venereal disease in the United Kingdom. A second study. Br J Vener Dis 56: 6–11

Centers for Disease Control (1984) Syphilis – United States, 1983, MMWR 33: 433–441

Collart P, Bovel L-J, Durel P (1964) Significance of spiral organisms found after treatment in late human and experimental syphilis. Br J Vener Dis 40: 81–89

Department of Health and Social Security (1985) Return for the year ended 31 December 1984 made by physicians in charge of treatment centres for sexually transmitted diseases (Form SBH60)

Drusin LM, Homan WP, Dineen P (1976) The role of surgery in primary syphilis of the anus. Ann Surg 184: 65–67

Duncan WC, Knox JM, Wende RD (1974) The FTA-ABs test in dark-field positive primary syphilis. JAMA 228: 859–860

Dunlop EMC (1985) Survival of treponemes after treatment: comments, clinical conclusions and recommendations. Genitourin Med 61: 293–301

Feher J, Somogyi T, Timmer M, Józsa L (1975) Early syphilitic hepatitis. Lancet ii: 896–899

Felman YM (1983) Syphilis. In: Ostrow DG, Sandholzer TA, Felman YM (eds) Sexually transmitted diseases in homosexual men. Plenum Medical Book Company, New York and London, pp 37–56

Geer LL, Warshauer DM, Delaney DJ (1985) Pulmonary nodule in secondary syphilis. Australas Radiol 29: 240–242

Gjestland T (1955) The Oslo study of untreated syphilis: an epidemiologic investigation of the natural course of syphilitic infection based on a re-study of the Boeck-Bruusgaard material. Acta Dermatol Venereol 35 [Suppl 34]: 11–368

Johns DR, Tierney M, Felsenstein D (1987) Alteration in the natural history of neurosyphilis by concurrent infection with the human immunodeficiency virus. N Engl J Med 316: 1569–1572

Johnson MH, Thompson EJ (1986) Diagnosis of neurosyphilis. Hospital Update 12: 561–563

Judson FN (1977) Sexually transmitted diseases in gay men. Sex Transm Dis 4: 76–78

Judson FN (1986) Infectious syphilis primary, secondary and early latent. In: Felman YM (ed) Sexually transmitted diseases. Churchill Livingstone, New York, Edinburgh, London, Melbourne, pp 23–37

Keisler DS, Starke W, Looney DJ (1982) Early syphilis with liver involvement. JAMA 247: 1999–2000

King A, Nicol CS (1975) Venereal diseases, 3rd edn. Bailliere Tindall, London

Knox JM, Rudolph AH (1984) Acquired infectious syphilis. In: Holmes KK, Mardh P-A, Sparling PF, Wiesner PJ (eds) Sexually transmitted diseases. McGraw-Hill, New York, pp 305–313

McDermot W (1958) Microbial persistence. Yale J Biol Med 30: 257–291

McMillan A, Lee FD (1981) Sigmoidoscopic appearance of the rectal mucosa in homosexual men. Gut 22: 1035–1041

Mohr JA, Griffiths W, Jackson R, Saadah H, Bird P, Riddle J (1976) Neurosyphilis and penicillin levels in cerebrospinal fluid. JAMA 236: 2208–2209

Ovčinnikov NM, Korbut SE, Bednova VN, Timčenko GF, Milonova TI (1973) Long-term results of penicillin treatment of early and late forms of syphilis in the rabbit. Br J Vener Dis 49: 413–419

Peterson CS, Pederson NS (1983) The profile of early infectious syphilis in Denmark. Dan Med Bull 30: 49–51

Putkonen T, Salo OP, Mustakallio KK (1966) Febrile Herxheimer reaction in different phases of primary and secondary syphilis. Br J Vener Dis 42: 181–184

Quinn TC, Corey L, Chaffee RG, Schuffler MD, Brancato FP, Holmes KK (1981) The etiology of anorectal infections in homosexual men. Am J Med 71: 395–406

Quinn TC, Lukehart SA, Goodell S, Mkrtichian E, Schuffler MD, Holmes KK (1982) Rectal mass caused by *Treponema pallidum*: confirmation by immunofluorescent staining. Gastroenterology 82: 135–139

Rockwell DH, Yobs AR, Moore MB (1964) The Tuskagee study of untreated syphilis. Arch Intern Med
    114: 792–798
Rosahn PD (1947) Autopsy studies in syphilis. J Vener Dis Inform [suppl 21]. U S Public Health Service,
    Venereal Disease Division
Sachar DB, Klein RS, Swerdlow F, Bottone E, Khilnani MJ, Waye JD, Wisniewski M (1974) Erosive
    syphilitic gastritis: dark-field and immunofluorescent diagnosis from biopsy specimen. Ann Intern
    Med 80: 512–515
Schober PC, Gabriel G, White P, Felton WF, Thin RN (1983) How infectious is syphilis? Br J Vener Dis
    59: 217–219
Schroeter AL, Turner RH, Lucas JB, Brown WJ (1971) Therapy for incubating syphilis. JAMA 218: 711–
    713
Short DH, Knox JM, Glicksman J (1966) Neurosyphilis, the search for adequate treatment. Arch Der-
    matol 93: 87–91
Smego RA, Moreadith RW, Kleist PC, Granger DL (1984) Secondary syphilis masquerading as AIDS in
    a young gay male. North Carolina Med J 45: 253–254
Smith CA, Kamp M, Olansky S, Price EV (1956) Benzathine penicillin G in the treatment of syphilis. Bull
    WHO 15: 1087–1096
Spangler AS, Jackson JH, Fiumara JJ, Warthin TA (1964) Syphilis with a negative blood test reaction.
    JAMA 189: 87–90
Stokes JH, Beerman H, Ingraham NR (1945) Modern clinical syphilology, 3rd edn. Saunders, Philadel-
    phia
Tramont EC (1976) Persistence of Treponema pallidum following penicillin G therapy. Report of two
    cases. JAMA 236: 2206–2207
von Krogh G, Hellström L, Bottiger M (1986). Declining incidence of syphilis among homosexual men in
    Stockholm. Lancet ii: 920–921
Wende RD, Mudd RL, Knox JM, Holder WR (1971). The VDRL slide test – 322 cases of dark-field
    positive primary syphilis. South Med J 64: 633–634
Willcox RR, Goodwin PG (1971) Nerve deafness in early syphilis. Br J Vener Dis 47: 401–406

Chapter 8

AIDS: Epidemiology and Clinical Aspects

M. W. Adler and I. V. D. Weller

Epidemiology

A "New" Disease

In the summer of 1981, reports of five cases of *Pneumocystis carinii* pneumonia (PCP) and 26 of Kaposi's sarcoma (KS) were reported amongst homosexual men in Los Angeles, California and New York (Gottlieb et al. 1981a,b; Friedman-Kien et al. 1981, 1982; Hymes et al. 1981). This complex of diseases was subsequently termed the acquired immune deficiency syndrome (AIDS). In the United States, the Centers for Disease Control (1982) originally defined a sufferer from the disease as a person:

1. With a reliably diagnosed disease that is at least moderately indicative of an underlying cellular immune deficiency; for example, Kaposi's sarcoma in a patient aged less than 60 years, or opportunistic infection.
2. Who has no known underlying cause of cellular immune deficiency nor any other cause of reduced resistance reported to be associated with the disease.

This definition was accepted for surveillance purposes by most other countries and was the one used within the UK. The definition was subsequently slightly modified in June 1985 in the light of laboratory tests to detect antibody and included additional serious conditions (Centers for Disease Control 1985a). A further change took place in September 1987. A new definition now includes human immunodeficiency virus (HIV) encephalopathy, HIV wasting syndrome and a wider range of diseases indicative of AIDS (Centers for Disease Control 1987). Also, in certain circumstances it permits the inclusion of cases without evidence of HIV infection or with laboratory evidence against HIV infection. The aetiological agent associated with AIDS has been known by various names since first discovered in 1983 – human T lymphotropic virus type III (HTLVIII), lymphadenopathy associated virus (LAV) and AIDS-associated virus (ARV) (Gallo et al. 1983; Barre-Sinoussi et al. 1983). In 1986 a subcommittee of the International Committee for the Taxonomy of Viruses (Biber-feld et al. 1987) suggested that the generic name for the virus should be the human immunodeficiency virus (HIV).

It is now realised that infection with HIV can manifest itself in many ways, from asymptomatic infection to severe immunodeficiency, opportunistic infections and other cancers (for example non-Hodgkin's lymphoma (Ziegler et al. 1982)). In view of this, a further classification system was derived in May 1986, which took account of the natural history and variable presentation of HIV infection (Centers for Disease Control 1986b) (Table 8.1).

It is now also appreciated that groups other than homosexual men are at risk of AIDS. These include intravenous drug abusers, haemophiliacs and heterosexuals. Haitian immigrants were also included early on but are no longer considered to be a high-risk group.

Table 8.1. Summary of classification system for human immunodeficiency virus

Group I	:	Acute infection
Group II	:	Asymptomatic infection
Group III	:	Persistent generalised lymphadenopathy
Group IV	:	Other disease
Subgroup A	:	*Constitutional disease* (fever > 1 month, weight loss > 10% base line, diarrhoea > 1 month)
Subgroup B	:	*Neurological disease* (dementia, myelopathy, peripheral neuropathy)
Subgroup C	:	*Secondary infectious diseases* C1 Those specified in CDC surveillance definition C2 Others: oral hairy leucoplakia, multidermatomal, herpes zoster, recurrent *Salmonella* bacteraemia, nocardiosis, tuberculosis, oral *Candida*
Subgroup D	:	*Secondary cancers* (Kaposi's sarcoma, non-Hodgkin's lymphoma, primary cerebral lymphoma)
Subgroup E	:	*Other conditions*

From Centers for Disease Control (CDC) 1986b.

Table 8.2. AIDS – adult patient groups – USA and UK, end December 1987

Patient groups	USA		UK	
	N	%	N	%
Homosexual men	31 825	(65)	1032	(85)
Intravenous drug user	8411	(17)	19	(1.5)
Homosexual male and i.v. drug user	3689	(8)	19	(1.5)
Haemophilia	484	(1)	70	(6)
Received blood	1124	(2)	24	(2)
Heterosexual contact	1964	(4)	44	(4)
Other/miscellaneous	1509	(3)	6	(0.5)
Total	49 006	(100)	1214	(100)
	M :	93%	M :	97%
	F :	7%	F :	4%

i.v., intravenous; N, number of patients; M, male; F, female.

Data from Communicable Disease Surveillance Centre, Communicable Disease Report, CDR Weekly Edition 88/3; and Centers for Disease Control, AIDS Weekly Surveillance Report – United States AIDS program, Center for Infectious Diseases, December 1987.

Table 8.3. AIDS cases, by date of report and doubling time,
USA, 1981 through December 1986

Cumulative cases reported	Date		Doubling time (months)
110	September	1981	—
220	January	1982	5
439	June	1982	6
878	December	1982	6
1756	July	1983	7
3512	February	1984	8
7025	December	1984	9
14 049	October	1985	11
28 098	December	1986	13

Data from Centers for Disease Control 1986c.

Table 8.4. Years of potential life lost (YPLL) before age 65 years and
changes in YPLL, New York City, 1984

	YPLL	Change in YPLL from 1982 (%)
Homocide and suicide	47 900	−14
Heart disease	41 600	−2
Malignant neoplasm	33 900	+5
AIDS	24 400	+510
Chronic liver disease and cirrhosis	18 700	−3
Accidents	11 600	−19
Pneumonia and influenza	8900	+86
Cerebrovascular disease	5800	+14
All causes	252 100	+8

Data from Centers for Disease Control 1985c.

Some 49 006 adult cases have been reported in the USA (December 1987), 8839 in Europe (December 1987) and 1214 in the UK (December 1987) (Table 8.2). In the UK, proportionately more homosexuals have been notified than in America, 86% of cases compared with 73%. The doubling time of cases in the USA has slowed down since 1981 and 1982. Thus, the actual number of cases reported for every six-month period increases, but not in an exponential fashion, as can be seen from the lengthening of the doubling time (Table 8.3). The overall and eventual fatality rate are 46% and virtually 100% respectively. The proportion of patients in the United States who are now dead is 84% for those notified in 1981 and 31% for those notified in 1984; equivalent figures for the UK are 75% and 30%. Median survival after diagnosis varies with the manifestation of the disease and is approximately nine months for patients with PCP and 31 months for those suffering from KS (Rivin et al. 1984). In the UK, the figures are similar, with a mean survival of 12.5 months for patients with PCP and 21.2 months with KS (Marasca and McEvoy 1986).

Within the USA, the rates for cases of AIDS per million of the population show wide geographical variation. New York has a rate of 991 per million, San Francisco 966, Miami 584, Newark 393, Los Angeles 363, compared with 140 per million for the USA as a whole. Likewise in some cities such as New York, AIDS is the major cause of premature death in young men. Table 8.4 indicates the changes in years of potential life lost between 1982 and 1984 in that city. Over this period, AIDS as a contributor has increased by over 500%, a six times greater increase than the next commonest increasing diseases, pneumonia and influenza.

It is reassuring that the epidemiology of the groups at risk of AIDS and HIV infection in the UK and USA has remained reasonably constant. Thus, transmission of the virus is reported by the following routes – sexual (predominantly homosexual), blood and blood products, needle stick injury, via contaminated needles in drug addicts, and vertically from mother to child. The American data show that the proportion of cases belonging to the different risk groups has not altered substantially from the early (854 cases before December 1982) and the later ones (3830 cases from December 1983 to November 1984) (Centers for Disease Control 1984).

The miscellaneous group born in America, which does not fall into any recognised "risk group", remains proportionally small and reasonably constant at between 2.9% and 4.1% of cases from 1982 to 1986 (Centers for Disease Control 1986d). Some of this group may not in fact have AIDS and some may have had intercourse with an AIDS sufferer or anti-HIV-positive person without realising this at the time. It is possible, but highly unlikely, that some of those in the miscellaneous group were social and family contacts who contracted HIV infection through a non-sexual route. Finally, some may be homosexual or from another high-risk group and unwilling to admit it. The Centers for Disease Control and local epidemiologists investigate all remaining cases that do not fall into the recognised groups and are classified in the first instance as "undetermined". As a result of this, 72% of such cases have been reclassified after risk factors were subsequently identified or the patient did not in fact have AIDS.

Females, unless they belong to the risk group of intravenous drug abusers and/or prostitutes, are not usually at high risk of acquiring AIDS in Europe or the United States. It is suggested that some of the cases in America that do not belong to any risk group may have contracted AIDS through heterosexual contact with infected prostitutes (Chamberland et al. 1984). By November 1984, 263 (3.8%) of AIDS cases in America belonged to the "non-characteristic" group (Centers for Disease Control 1984). Detailed investigations of 65 of the males in this group indicated that 17 (26%) admitted to sexual intercourse with female prostitutes. Five of the 17 admitted to over 100 heterosexual partners in the previous five years. One of the nine women belonging to the non-characteristic group was a former prostitute. Another study has shown that, of 36 male AIDS cases without known risk factors in New York, 11 reported one or more episodes of intercourse with female prostitutes in the five years prior to the onset of the disease and two further men had experienced intercourse with intravenous-drug-abusing women (R. R. Abkin, A. Lekatsas and J. Walker, unpublished work presented at the International Conference on Acquired Immunodeficiency Syndrome (AIDS), Atlanta, GA, 14–17 April 1985). Contact with prostitutes is also cited as a mode of transmission in another study in army personnel (R. R. Redfield, P. D. Rhkam and J. Z. Salahuddin, unpublished work presented at the International Conference on Acquired Immunodeficiency Syndrome (AIDS), Atlanta, GA, 14–17 April 1985); however, because of strict penalties against homosexuality and drug abuse in the army, homosexual as opposed to prostitute contact cannot be ruled out as a risk factor.

Some women engaging in prostitution do so to finance their drug addiction and, thus, also belong to a high-risk group for AIDS. In the United States, approximately a third of women who had sought treatment for their drug addiction had practised prostitution to obtain money for drugs (Grizburg et al. 1988). Finally, the largest number of female cases in the USA (1815; 50%) occurred in intravenous drug users. There are only 19 cases of AIDS in intravenous drug abusers in the UK (14 male, five female – December 1987); no substantiated cases have occurred in prostitutes.

Despite the current low incidence of HIV infection and AIDS in the UK amongst prostitutes, it is important to monitor any changes that may occur (Barton et al. 1985).

HIV seropositivity is very low among heterosexual men and women not belonging to the established risk groups (Sarngadharan et al. 1984; S. J. Weiss et al. 1985). However, heterosexual transmission of AIDS and HIV is documented in the United States, usually from men to women (Centers for Disease Control 1983; Harris et al. 1983). Limited studies have shown that the risk of transmission of infection from seropositive husbands to their wives has ranged from 2 out of 21 (10%) to 5 out of 7 (71%) (Kreiss et al. 1985; Redfield et al. 1985). By December 1987, 2.7% (33 cases) of AIDS in adults had occurred in females in the UK and in the United States 3601 out of a total of 49 006 (7.0%) adult cases. Of these American women, 29% had contracted AIDS through heterosexual sexual contact, the commonest risk factor still being intravenous drug abuse (50%).

In Africa, heterosexual transmission appears to be by far the most common mode of transmission of HIV. The male to female ratio of cases of AIDS is virtually one to one and a case control study showed that male cases were more likely than controls to be promiscuous and have had intercourse with prostitutes. The prevalence of antibody in some prostitutes is reported to be as high as 88% (Piot et al. 1984; Van de Perre et al. 1984, 1985; Clumeck et al. 1985). In Kenya, the prevalence in this risk group was 59% in 1985/86 having risen from 4% in 1980/81.

Changing Prevalence

The advent of an effective antibody test has allowed for a clearer understanding of the changing prevalence and natural history of HIV infection. The availability of an antibody test and access to stored samples of sera collected for other reasons from groups of homosexual men has given a clear indication of how the epidemic of HIV infection has developed. In San Francisco, one of the cities both seeing the epidemic early and caring for a disproportionate number of cases, it has been possible to describe the unfolding epidemiology in detail (Curran et al. 1985). Thus, in 1978 in a cohort of homosexual men, the proportion who were anti-HIV positive was 4%. By 1980 this had already increased six-fold, and it was only when the seroprevalence had reached 24% that cases of AIDS started to occur (Table 8.5). Also it has become clear from this that the number of individuals infected with HIV, and therefore antibody positive, is always greater than the total number of cases. In 1983 in San Francisco there were 46-fold more infected individuals than cases and by 1984 it was 28-fold.

A similar picture has been seen amongst homosexual men in other large European cities. For example, amongst homosexual men attending a Central London clinic for sexually transmitted diseases (STD clinic), the prevalence of antibody to HIV has risen from 3.7% in March 1982 to 21% in July 1984 (Fig. 8.1) (Carne et al. 1985b). These data confirm that the proportion of individuals infected needs to be high before cases become apparent, but it also underlines the importance of early intervention in the epidemic process when the seroprevalence is low. Once cases start to appear, the epidemic drives itself and a much greater effort in terms of control and medical care is then required. One of the tragedies surrounding AIDS is that this has never been realised and continues to be ignored. For example, in the UK the sero-prevalence in homosexual men attending STD clinics in the provinces was low in 1984 at 5.1% increasing to 11.0 by 1985 (Table 8.6) (Mortimer et al. 1985a; Carne et al. 1985b; Jesson et al. 1985). The opportunity for intervention through an extensive health education programme has been lost. It was not until March 1986 that the British Government started any health education initiatives. The same rapid increase in prevalence has been seen in other high-risk groups such as intravenous-drug abusers (Table 8.7). In Bari, Southern Italy, the prevalence rose from 6.0% to 76% in the space of only six years and in Switzerland doubled in four years (Mortimer et al. 1985b; Angarano et al. 1985).

Table 8.5. Estimate of number of individuals with HIV antibody and AIDS, 1978–1984 (from San Francisco CDC Cohort Study ($n = 6875$))

Variable	1978	1979	1980	1981	1982	1983	1984
Seropositive (%)	4	12	24	35[a]	46[a]	57[a]	68
Estimated number seropositive	275	825	1650	2406	3162	3919	4675
Cumulative number reported with AIDS	0	0	2	14	41	84	166

CDC, Centers for Disease Communication; n, number of patients.
From Curran et al. 1985.
[a]Estimated.

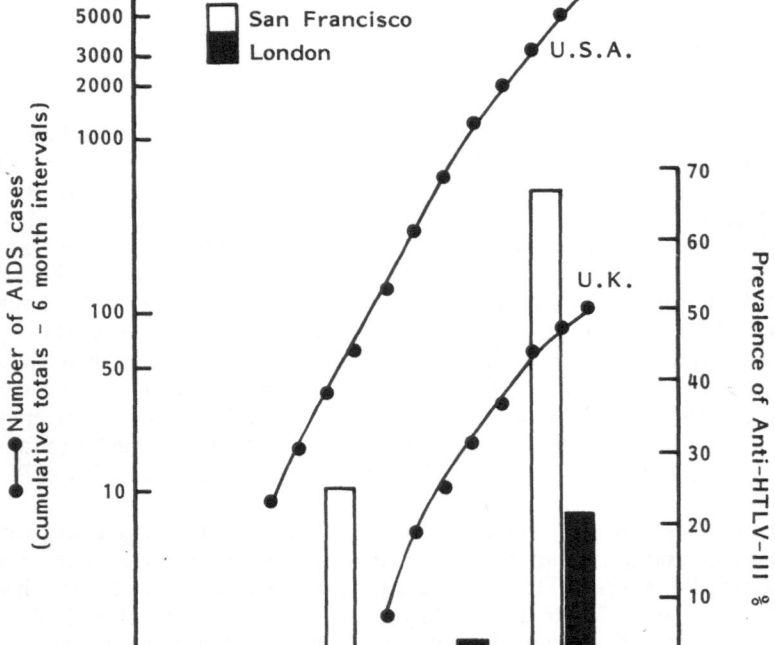

Fig. 8.1. Rise in prevalence of AIDS cases and anti-HIV in the *USA* and the *UK* (from Carne et al. 1985b).

Table 8.6. Prevalence of anti-HTLVIII in homosexual men attending STD clinics

Location	Year	Specimens	Prevalence (%)
London	1982	107	3.7
	1984	124	21.0
	1985	432	35.0
5 provincial	1984	955	5.1
centres	1985	—	11.0

Data from: Carne et al. 1985b; Mortimer et al. 1985a; Jesson et al. 1985.

Table 8.7. Increasing prevalence (%) of anti-HTLVIII in intravenous drug addicts, Europe

Year	Prevalence in:			
	Italy	Switzerland	England	Scotland
1980	6.0			
1981	—			
1982	—	16.0		
1983	—			
1984	—		2.5	
1985	76.0	32.0	10.0	60.0

Data from: Angarano et al. 1985; Mortimer et al. 1985b.

Natural History of Infection

Series of cohort studies give some indication of the natural history of infection with HIV. Goedert et al. (1986) analysed the rate of progression from infection with HIV to AIDS in five different groups over three years. This actuarial analysis (Table 8.8) showed a rate of progression ranging from 8% to 34% in Danish and New York homosexual men, respectively, with a high rate also in New York intravenous-drug abusers (25%). The differences in the rates of progression cannot be explained easily but it is possible that the New York men had in fact been anti-HIV positive for a longer period prior to entry into the study and exposed to more potent cofactors than the other groups. It has to be stressed that the epidemiology of infection with HIV is continuously unfolding and therefore changing. Some scientists feel that the proportion of individuals infected who will eventually develop AIDS is bound to increase. The same is the case for the chronic manifestations of infection such as persistent generalised lymphadenopathy (PGL). Cohort studies indicate a varying proportion with this condition progressing to full-blown AIDS (Table 8.9) (Mathur-Wagh et al. 1984, 1985; Metroka et al. 1983; Abrams et al. 1985). Again, it is clear that the longer individuals are followed the greater the proportion progressing from PGL to AIDS. Thus, Mathur-Wagh et al. (1985) reported 19% of their patients progressed to AIDS at a median follow up of 22 months but that, when this time had increased, the proportion rose to 29%. Again, this confirms that time is one of if not *the* most potent factor in progression. Some of the clinical parameters that help the clinician to predict such progression are discussed later in the chapter.

Table 8.8. Three-year incidence (%) of AIDS in five cohorts

Cohort		3-year incidence (%)	
Homosexual men	(New York)	34.2	(±8.0)
Homosexual men	(Washington DC)	17.2	(±6.4)
Homosexual men	(Copenhagen/Aarhus)	8.0	(±5.5)
Haemophiliacs	(Pennsylvania)	12.8	(±4.8)
Intravenous-drug abusers	(New York)	25.0	(±15.3)

Numbers in parentheses indicate standard error of the mean.
From Goedert et al. 1986.

Table 8.9. PGL prospective studies

No. of patients	Follow-up (months)	AIDS		References
		(N)	(%)	
42	Median 22	8	(19)	Mathur-Wagh 1984
42	Longest 60	12	(29)	Mathur-Wagh 1985
90	8 to 19	15	(17)	Metroka et al. 1985
70	Longest 16	0	(0)	Abrams et al. 1985
200	Longest 61	16	(8)	Abrams et al. 1985
100	Median 24 (9–50)	13	(13)	Carne et al. 1987a

Prevalence and Control

Since no cure or vaccine is currently available, the cornerstone of control currently is prevention. Good and accurate health education is required for those at both low and high risk. Homosexual men have been advised to avoid donating blood, organs and semen and to modify their sexual behaviour to avoid practices that are particularly unsafe in transmitting the virus (for example, anal intercourse) (Table 8.10). Essentially, penetrative intercourse carries with it the highest risk of infection, particularly if one is the receptive partner. Evidence is beginning to be collected that indicates that certain of the high-risk homosexual men are altering their sexual behaviour in terms of both reducing the number of casual partners and adopting safer sex practices.

Evidence for such changes comes from indirect measures such as changing patterns of STDs and detailed studies of sexual practices. In the USA and the UK changes in the rates of gonorrhoea and syphilis have been reported in homosexual men attending clinics (Judson 1983; Weller et al. 1984; Carne et al. (1987b). In London, the rate of gonorrhoea had fallen from 15.0% in 1982 to 5% by 1985 (Fig. 8.2). Changing sexual practices have been reported in homosexual men between August 1984 and April 1985 (Centers for Disease Control 1985b). The survey was carried out by random telephone interviewing and indicated that more men had entered a monogamous or celibate relationship over this time interval and fewer were having more than one partner (Table 8.11). Changes in sexual practices have also been observed in a cohort of 100 men in London (Table 8.12). They have shown a reduction in the number of partners and trends towards safer sex practices. Condom usage had increased but not to a statistically significant degree.

Table 8.10. AIDS safer-sex guidelines

No risk	*Low risk*
Solo masturbation	Mutual masturbation
Massage away from the genital area	Dry kissing
	Body rubbing
Medium risk	*High risk*
Wet kissing	Anal and vaginal sex may be safe if a condom is used
Fellatio (sucking) may be safe if you stop before climax	Fisting (insertion of hand/fist into rectum)
Urination (water sports) external only	Sharing sex toys and needles
Anilingus (rimming)	Any sex act that draws blood

Risk increases with multiple partners

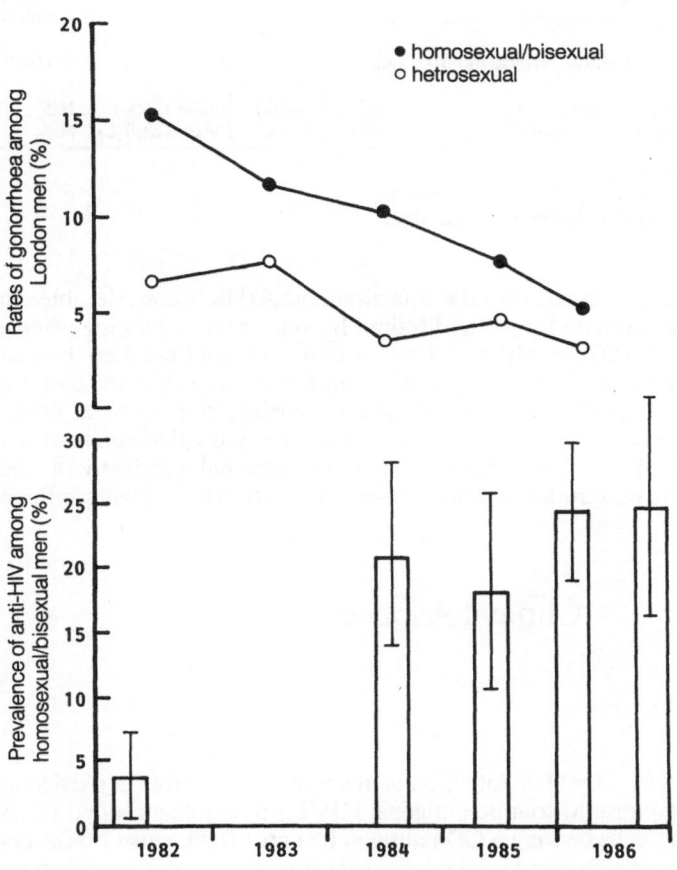

Fig. 8.2 Prevalence of anti-HIV and gonorrhoea rates *homosexual/bisexual* and *heterosexual* London men (from Carne et al. 1987b).

Table 8.11. Changes in selected self-reported sexual practices among gay and bisexual men – San Francisco, CA, August 1984 and April 1985

	August 1984 (%)	April 1985 (%)
Monogamous, celibate or no unsafe sexual activity outside a primary relationship	69	81
More than one sexual partner during last 30 days	49	36
Anal intercourse (without a condom) with secondary partner during past 30 days	18	12
Oral sex (with exchange of semen) with secondary partners during past 30 days	17	7

Data from Centers for Disease Control 1985b.

Table 8.12. Behaviour change in 100 HIV study cohort members

	1984/85		1986		P value
Partners per month; median (range)	3	(0–40)	1	(0–45)	<0.0001
One-night partners sometimes (number men)	97		64		<0.0005
Active anal SI with 2 or more partners in a typical month (number men)	27		8		<0.0005
Passive anal SI with 2 or more partners in a typical month (number men)	41		16		<0.0005
No. of men always using a sheath in active anal SI	9/67[a]	(13.4%)	15/60[b]	(25%)	N.S.
No. whose partners always use sheath in anal SI	8/67[a]	(11.9%)	15/63[b]	(23.8%)	N.S.

SI, sexual intercourse; NS, not significant.
[a]33 men reported no active anal SI and 33 no passive anal SI.
[b]40 men reported no active anal SI and 37 no passive anal SI.
From Carne et al. 1987b.

The cost of caring for patients with HIV infection and AIDS is considerable. In New York it has been estimated that the lifetime hospital cost of looking after a patient with AIDS is $134 000 (Hardy et al. 1985), whereas in San Francisco the cost is less at between $25 000 and $32 000 (Skitovsky et al. 1986). In the UK the cost in a Central London health authority has been calculated at £6800 (Johnson et al. 1986). These costs take into account only hospital costs, and clearly if calculations of the time off work, effect of death of young persons on national productivity and domiciliary services were undertaken it would be seen that AIDS is, and will continue to be, very costly.

Clinical Aspects

Introduction

The primary defect induced by HIV infection is in cellular immunity. In particular the cellular immune response to soluble antigens. HIV1 is a double-stranded RNA retrovirus, which infects cells bearing a CD4 antigen receptor (CD, cluster differentiation). The CD4 antigen expressed by the helper/inducer subset of T lymphocytes makes them most susceptible to infection. These lymphocytes have a pivotal role in the immune response. Their depletion and/or a functional abnormality that results from HIV infection leads not only to a loss of cytotoxic T cell responses but also a reduction of natural killer cell and monocyte function and B cell abnormalities with polyclonal B cell activation, hypergammaglobulinaemia and impairment of specific antibody responses to new antigens (Lawrence 1985). In addition, HIV has been shown to infect macrophages, B lymphocyte cell lines and cells in the central nervous system (Gartner et al. 1986). The antibodies produced to HIV core and envelope proteins in vivo are a readily detectable marker of exposure. However, they have only weak in-vitro neutralising activity relative to neutralising antibodies to other human retroviruses. They are present in latent infection, and during productive viral replication, and there appears to be no relationship between titre and disease expression (R. A. Weiss et al. 1985).

Isolates of HIV show considerable genetic heterogeneity, particularly in the envelope gene region (Wong-Staal and Gallo 1985). Recently a similar virus (HIV2) has been isolated from patients with AIDS from West Africa, and two caucasian homosexual men in Paris. Its morphological and biological features are similar to HIV1 but it differs in some of its antigenic components, notably the envelope glycoprotein (Clavel et al. 1986; Brucker et al. 1987).

Acute Infection (Centers of Disease Control (1986b), Stage I)

Although most acute infections are asymptomatic, a number of syndromes have been described. The commonest is a non-specific viral or glandular-fever-type illness occurring one to six weeks after infection. This may present with fever, sweats, malaise, myalgia, arthralgia, sore throat, nausea, headache, diarrhoea, photophobia, a rash and lymphadenopathy (Cooper et al. 1985). An acute meningo-encephalitis myelopathy and polyneuropathy have also been described (Carne et al. 1985a; Denning et al. 1987). The encephalitis may present with a prodrome of malaise, fever and personality change, progressing to severe headache, convulsions and various degrees of coma. Seroconversion for anti-HIV occurs typically 4–12 weeks after acute infection, and in the acute illness may not develop until after the symptoms have resolved. Longer delays in the development of anti-HIV have been described. The inner-core polypeptide (P24), of molecular weight 24 000, appears transiently before the antibody in the acute infection (Fig. 8.3). It reappears in the later stages of infection together with loss of anti-P24 and both appear to be markers of disease progression. In addition to P24, an assay for IgM antibody may also fill in the early serological window of infection.

Chronic HIV Infection (Centers for Disease Control (1986b), Stages II, III and IVa)

The commonest finding in chronic HIV infection is generalised lymphadenopathy. About a third of anti-HIV positive patients fulfil the Centers for Disease Control definition of persistent generalised lymphadenopathy (PGL) with nodes of at least 1 cm in diameter in two extrainguinal sites for three months or more. Many others have lesser degrees of lymphadenopathy. In both categories, the nodes are symmetrical, mobile and non-tender. It was thought that asymptomatic patients with PGL had a

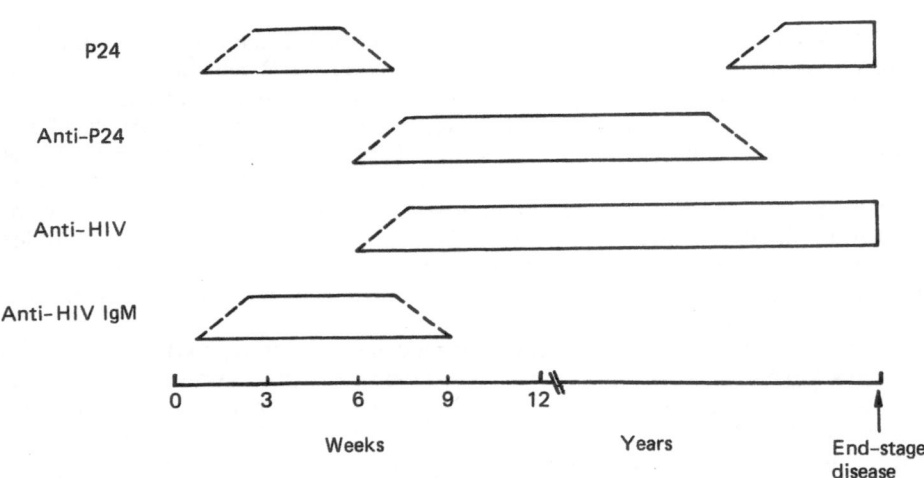

Fig. 8.3. Serological markers of acute and chronic infection.

more advanced infection than those without. However, a prospective study has shown that asymptomatic anti-HIV-positive patients, with and without PGL, progress to AIDS at the same rate (Polk et al. 1987). Referral of every patient with lymphadenopathy for lymph node biopsy is inappropriate. Biopsy is unrewarding in asymptomatic patients, revealing only non-specific follicular hyperplasia. If there are constitutional symptoms such as weight loss or fever, or the nodes are markedly asymmetrical, painful or rapidly enlarging or if there is an extranodal mass or coexisting hilar nodes, then biopsy is essential (Rashleigh-Belcher et al. 1986). This will exclude tumours such as lymphoma or lymphadenopathic Kaposi's sarcoma (KS) or an opportunistic infection such as mycobacteria. Tiredness, lethargy, fever, night sweats and weight loss may occur, without any other identifiable cause, and presumably are a direct effect of chronic HIV infection.

Minor opportunistic infection is common and the skin and mucous membranes are frequently affected (Table 8.13). Skin conditions encountered include a range of infections: dermatophytic fungal, viral (recalcitrant warts, molluscum contagiosum, herpes simplex and zoster) and bacterial (folliculitis, impetigo and furunculosis). The spectrum of other skin disorders range from mere dryness to severe seborrhoeic dermatitis. The recently described "hairy" leucoplakia appears typically as greyish-white lesions on the lateral borders of the tongue (Greenspan et al. 1985). Its histological appearance is characteristic, with keratin projections resembling hairs, koilocytosis and mild atypia. Epstein–Barr virus (EBV) DNA has been demonstrated in these lesions by hybridisation with specific probes and there have been anecdotal reports of response to acyclovir (Friedman-Kien 1986).

The risk of progression to AIDS from published studies ranges from 7% to 30% over three to four years (Goedert et al. 1986). However, prospective cohort studies only represent a short period of observation from which to draw conclusions about the natural history of HIV infection. Prognostic markers for disease progression are also being sought from these studies (Anonymous 1986) (Table 8.14). Clinical features such as constitutional symptoms, oral *Candida*, leucoplakia, herpes zoster and disappearance of lymph nodes have been recorded. Haematological abnormalities encountered include anaemia, a raised erythrocyte sedimentation rate and

Table 8.13. Minor opportunistic conditions – skin + mucous membranes

Seborrhoeic dermatis	
Fungal:	*Tinea*
	Candida
	Pityriasis versicolor
Bacterial:	Acneiform folliculitis
	Impetigo
Viral:	Herpes simplex I and II
	Herpes zoster
	Leucoplakia ?Epstein–Barr virus
	Human papilloma virus
	Molluscum contagiosum

Table 8.14. Prognostic markers

Clinical:	Oral *Candida*
	Herpes zoster
	Leucoplakia
	Constitutional symptoms
Laboratory:	↓ CD4-positive lymphocytes
	Anaemia
	Cytopenias
	↑ ESR
	↑ β2 microglobulin
	↑ IgA
	↓ Gamma interferon
	↓ Mitogenic responses to PWM
	↓ Anti-p24
	↓ Anti-P17
	↑ P24

ESR, erythrocyte sedimentation rate; PWM, pokeweed mitogen.

cytopenias. The lymphopenia is largely accounted for by a progressive depletion of CD4-positive lymphocytes. Neutropenia is probably due in part to anti-neutrophil antibodies, but counts rarely drop below $1.0 \times 10^9/l$ (Murphy et al. 1987). However, it is possible that a partially compensated immune neutropenia may predispose to more severe neutropenia developing with high-dose intravenous cotrimoxazole used for the treatment of *Pneumocystis carinii* pneumonia (PCP), with chemotherapy or with antiviral therapy. An autoimmune thrombocytopenia also occurs, with anti-platelet antibodies demonstrated against a target antigen on platelets (Stricker et al. 1985). Spontaneous bleeding, with counts below $20 \times 10^9/l$ is unusual but does occur. Response to steroids is transient and not without the theoretical risk of potentiating an opportunistic infection. Splenectomy results in a more long-lasting improvement in 80% of cases (Walsh et al. 1985) and can be preceded by high-dose intravenous gammaglobulin therapy, which usually transiently restores the platelet count to over $100 \times 10^9/l$ in 5 to 7 days.

Other immunological parameters that have been associated with progression include: impairment of in-vitro gamma-interferon production, a raised β2 micro-globulin and impaired lymphocyte responses to pokeweed mitogen and perhaps specific cellular responses to HIV. More recently a decline in titre or disappearance of anti-P24 (anti-gag, the antibody to HIV inner-core protein) and anti-P17 (antibody to the outer-core protein) and the appearance of P24 core antigen in serum have been shown to herald the development of the full-blown syndrome (Lange et al. 1986,

1987; Weber et al. 1987a). It remains to be seen which of these simple clinical or laboratory abnormalities as opposed to the more sophisticated research tests are the best predictors of progression to AIDS. Most of these abnormalities are markers of more severe chronic HIV infection and not determinants. Time itself may be the most important determinant of disease progression.

AIDS (Centers for Disease Control (1986b) Stage IVb–e)

Pulmonary Complications

PCP is the commonest life-threatening opportunistic infection in patients who progress from chronic HIV infection to AIDS. The presentation is subacute with malaise, fatigue, weight loss, and shortness of breath often developing over several weeks. Typical retrosternal or subcostal chest discomfort associated with increasing shortness of breath, a dry cough and fever finally cause the patient to seek help. The chest X-ray at presentation may be normal or show bilateral, fine infiltrates, typically perihilar (Fig. 8.4). The arterial oxygen tension is usually depressed, and the carbon monoxide transfer factor, where available, is low and may be the earliest detectable abnormality. The diagnosis is confirmed by cytological examination of induced sputum or by fibre-optic bronchoscopy, with bronchial lavage and transbronchial biopsy. At the same time, other causes of pneumonia or coexistent infection such as cytomegalovirus (CMV), mycobacteria and fungi may be excluded (Murray et al. 1984).

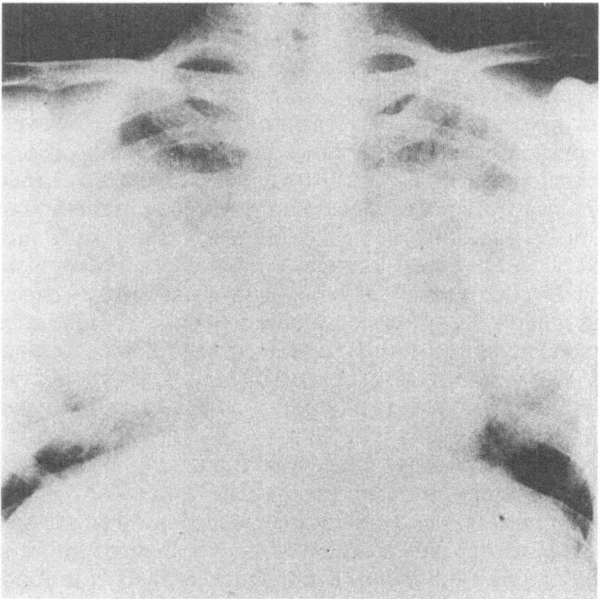

Fig. 8.4. Bilateral infiltrates of *Pneumocystis carinii* pneumonia (PCP)

Pyogenic bacterial causes of pneumonia should always be considered, particularly as their presentation may be atypical. In one series, 10% of episodes of pneumonia in AIDS patients were due to bacteria such as *Streptococcus pneumonia*, *Haemophilus influenzae*, *Branhamella* and group B streptococci (Polsky et al. 1986). The radiological appearances may include diffuse infiltrates as well as the more typical focal or lobar patterns. Pneumococcal pneumonia also occurs in anti-HIV-positive patients long before other manifestations of AIDS. Another cause of diffuse abnormality is lymphocytic interstitial pneumonitis, first described in paediatric AIDS and now recognised increasingly in adults. The cause is still uncertain, but it is certainly associated with HIV infection.

Infection with *Mycobacterium tuberculosis* may also occur. Pulmonary tuberculosis does not constitute a diagnosis of AIDS. A slowing of the downward trend of reported tuberculosis cases has occurred in the United States pari passu with the HIV epidemic (Centers for Disease Control 1986a). It seems that pulmonary tuberculosis appears in chronic HIV infection earlier than the atypical mycobacterial infections that complicate the severe immune depression of AIDS. The former responds to conventional anti-tuberculous therapy.

Gastrointestinal and Hepatic Complications (Table 8.15)

Retrosternal Discomfort: Dysphagia. Oral and oesophageal candidiasis was one of the features documented in the first cases of AIDS. It can be asymptomatic or cause oropharangeal discomfort and with oesophageal involvement, retrosternal chest pain and discomfort and/or difficulty on swallowing. Oral candidiasis alone does not fulfil the criteria for AIDS. Oesophageal involvement is best demonstrated by culture or biopsy at endoscopy, although plaques of *Candida* can often be demonstrated on barium swallow (Fig. 8.5). Ulceration may be focal or diffuse. Both cytomegalovirus and herpes simplex virus may cause a similar pattern of ulceration in the oesophagus (and also may affect the stomach and duodenum), and it may be difficult to differentiate them with barium studies. However, since oesophageal candidiasis is so common, a pragmatic approach in a patient with oral *Candida* and oesophageal symptoms would be to treat empirically for candidiasis and to only proceed to endoscopy should there be no response. One major drawback of this approach is that biopsy or culture-proven oesophageal candidiasis is required for the diagnosis of AIDS.

Diarrhoea, Malabsorption and Weight Loss. Diarrhoea is a common symptom of patients with chronic HIV infection, with and without other manifestations of AIDS. In many cases a cause is not found. Symptomatic treatment is all there is to offer. An enteropathy with villous atrophy and malabsorption has been described, but the prevalence of this condition and its aetiology, in particular when HIV is directly involved, have yet to be determined (Kotler et al. 1984).

Cryptosporidium is a coccidian protozoal parasite and probably the commonest pathogen isolated in AIDS patients with diarrhoea and certainly the commonest of the protozoal causes of diarrhoea, which also include *Isospora belli* and *Microsporidium* (DeHovitz et al. 1986; Dobbins and Weinstein 1985; Modigliani et al. 1985; see also Chap. 4). *Cryptosporidium* is widely distributed in the animal kingdom. In immunocompetent human hosts it produces a transient diarrhoeal illness. In HIV-infected hosts it can cause transient, intermittent or persistent diarrhoea, ranging

Table 8.15. Conditions affecting the gastrointestinal tract and liver in AIDS

Syndrome	Causes
Retrosternal discomfort/dysphagia	Candidiasis
	Cytomegalovirus (CMV)
	Herpes simplex (HSV)
Diarrhoea/weight loss/malabsorption	Unknown – enteropathy
	Cryptosporidium (*Isospora belli* and microsporidia)
	CMV/HSV
	Mycobacteria
	Enteric bacteria – *Salmonella, Campylobacter*
	Neoplasia
Hepatitis/cholestasis	Mycobacteria
	Cryptosporidium
	Cytomegalovirus
	Cryptococcus neoformans
	Drug induced
Neoplasia and miscellaneous	Kaposi's sarcoma
	Lymphoma
	Hairy leucoplakia
	Recalcitrant ano-rectal warts
	?Squamous oral/anal carcinoma

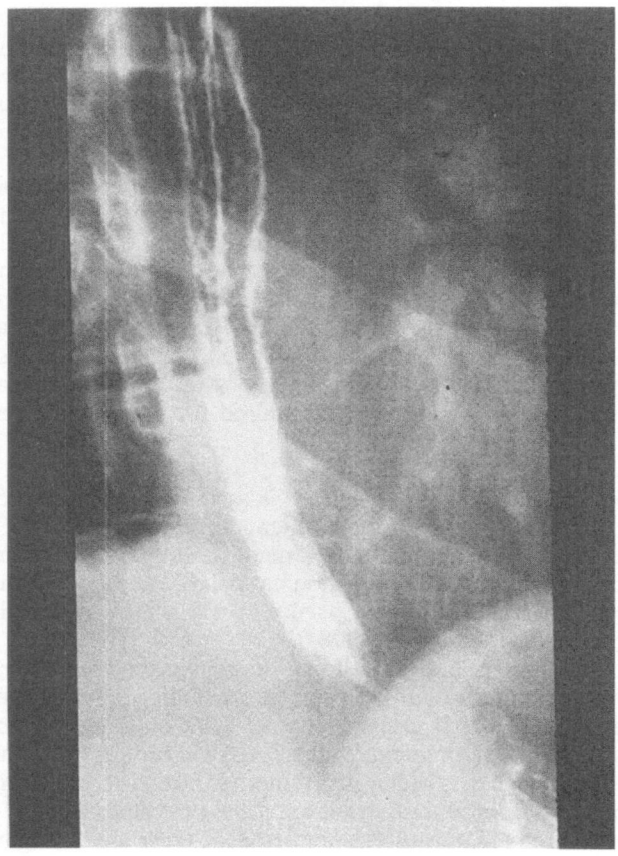

Fig. 8.5. Barium swallow oesophageal candidiasis (from Weller 1987).

from loose stool to watery diarrhoea, with colic and severe fluid and electrolyte loss. In reports before 1978, oocysts of *Cryptosporidium* were not identified in the stool and diagnosis rested on demonstrating the intermediate forms in biopsy material. However, in recent years improved staining techniques on stool specimens have allowed an easier diagnosis and have enhanced our understanding of the disease. The various forms of *Cryptosporidium* have been found in biopsy material from the pharynx to the rectum of infected humans, but infection seems to be most severe in the small bowel. The intermediate forms can be missed on histological examination of small-bowel biopsies. The organism does not appear to be invasive. There may be varying degrees of villous atrophy (Soave et al. 1984). Contrast studies are often normal, but may show a non-specific malabsorption pattern in the small bowel. Oocysts (4–5 μm) can be found in the stool. Direct smears of unconcentrated faecal samples may be all that is required with iodine or modified acid-fast stains. If these fail to reveal the oocysts, then concentration is required, such as a modified Sheather's flotation method followed by bright-field microscopy or staining (Ma and Soave 1983). The diagnosis should not be discounted without examining multiple specimens.

Two members of the herpesvirus group have been associated with gastrointestinal disease in AIDS patients; herpes simplex virus (HSV) and CMV (see also Chap. 3). These viruses have a number of common properties. Following primary infection, the viruses remain latent in their host. A state where the functional viral genome persists but there is no active replication. Reactivation with a high level of viral replication may occur at any time following primary infection and this may or may not be accompanied by clinical manifestations. Primary infection and reactivation of infection may cause severe disease in the immunocompromised. Serious disseminated infections occur in organ transplant recipients, with involvement of many organs, especially if primary infection occurs. The diagnosis rests largely with culture of the virus from the lesions. However, herpesviruses may be shed from a large number of epithelial surfaces in immunocompromised hosts and the presence of the virus may not necessarily indicate an aetiological link with the epithelial lesion.

In AIDS, CMV and HSV can cause focal or diffuse ulceration of the gut, from the mouth to the anus. Most commonly, HSV causes mucocutaneous lesions at the upper and the lower ends of the gastrointestinal tract, whilst CMV may mimic inflammatory bowel disease.

With progressive chronic HIV infection, recurrences of anogenital HSV may become increasingly frequent and/or severe. In AIDS, large, deep ulcers of the perianal area occur. Similar but smaller lesions may occur around the mouth. Perianal ulceration often occurs in the setting of a chronically ill patient, often with other opportunistic infections and severe weight loss and must be differentiated from pressure sores. The ulcers tend not to occur on pressure points and HSV is readily isolated from them. Ulceration may also occur in the oesophagus and bronchial tree (Macher 1984).

Gut involvement by CMV has been described in other immunosuppressed patients limited to isolated ulcerated segments, usually large bowel, with perforation or occasionally massive haemorrhage described. In AIDS, CMV has been associated with a syndrome that can mimic acute inflammatory bowel disease, with abdominal pain, fever and diarrhoea (Weber et al. 1987b). There may be diffuse or segmental ulceration (Balthazar et al. 1985). Toxic dilatation, perforation and haemorrhage have been described. Diagnosis is made by endoscopy with biopsy and culture. Histologically there is a non-specific inflammation, with dense, round (owl's-eye) intranuclear inclusion bodies in swollen cells. These inclusions are seen most readily in the vascular

endothelium of inflamed areas, suggesting that the colitis is caused by a virally induced vasculitis (Frager et al. 1986). Barium studies may reveal ulceration but may
not differentiate CMV infection from other causes of ulceration, such as HSV, KS
(Fig. 8.6) or lymphoma (see below). Disseminated CMV infection often occurs as a
terminal event in AIDS patients and there is no specific therapy of proven value.

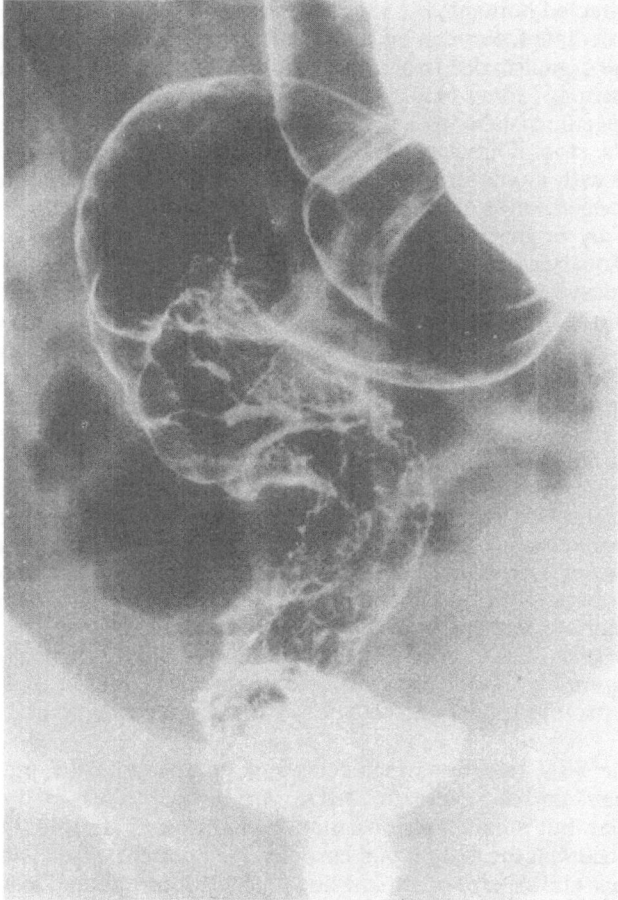

Fig. 8.6. Kaposi's sarcoma of rectum and sigmoid colon.

Atypical mycobacteria of the avium intracellulare complex are ubiquitous
organisms with little virulence for the immunocompetent host. Disseminated infection occurs in AIDS with multi-organ involvement. Gastrointestinal infection may
be associated with fever, weight loss, diarrhoea and malabsorption. Diagnosis can be
made by acid-fast staining of the stool and/or biopsy material or blood and tissue culture. Gut involvement may mimic Whipple's disease in appearance (Rotterdam and
Sommers 1985). The small bowel shows prominent folds, with periodic acid–Schiff-
positive foamy macrophages, containing the organisms, and filling the lamina propria. The bacteria are acid fast, unlike those of Whipple's disease. *Mycobacterium*

tuberculosis infection of the bowel does occur, but is less common. *Campylobacter* and *Salmonella* species infections may cause diarrhoea, but the latter more commonly present as a pyrexia of unknown origin with bacteraemia (Nadelman et al. 1985; Jarrett and Zeegen 1986). As with other infections in patients with AIDS, relapses are common following cessation of therapy.

Hepatitis and Cholestasis. Hepatitis in AIDS patients may present with fever, abdominal pain, hepatomegaly and abnormal liver function tests, in particular a raised alkaline phosphatase. In the absence of dilated bile ducts on ultrasound, needle biopsy most commonly demonstrates a granulomatous hepatitis, usually caused by atypical mycobacteria rather than *M. tuberculosis*. Atypical mycobacteria may be demonstrated on acid-fast staining or culture in the absence of granulomata. *Cryptococcus neoformans* may also be involved (Orenstein et al. 1985; Gordon et al. 1986). The herpesviruses may also occasionally cause hepatitis as part of a disseminated infection. With the multiple therapies being employed, a drug-induced hepatitis must always be considered in an AIDS patient with abnormal liver function tests (see Chap. 5).

More recently, acalculous cholecystitis and cholangitis have been described with an endoscopic retrograde cholangiographic picture similar to that of primary sclerosing cholangitis, with strictures and dilatation of the biliary tree (Fig. 8.7). Dilatation and irregularities of the pancreatic duct also have been reported. Histologically, there is non-specific inflammation and ulceration. *Cryptosporidium* and CMV have been demonstrated and/or isolated and implicated as a cause of this syndrome. Gram-negative bacteria and *Candida* have also been cultured (Margulis et al. 1986; Kavin et al. 1986; Cockerill et al. 1986; Gross et al. 1986).

The gut is commonly affected by the tumours of AIDS, namely KS and lymphoma (see below).

Neuropsychological Complications

Neurological disease occurs at all stages of HIV infection. The syndromes that have been reported in acute infection have been described. A chronic encephalopathy is a complication of chronic HIV infection characterised by a subacute dementia with motor dysfunction (Price et al. 1986; Navia et al. 1986a). Post-mortem studies suggest that up to 95% of AIDS patients will develop neurological involvement, while clinical assessment identifies at least two-thirds. This disorder may pre-date the diagnosis of AIDS and cases have been reported in whom death occurred due to neurological disease without the patient ever having had any other manifestation of AIDS. Early features are mild and may be described by both patient and friends as a general mental slowing, manifest as delayed motor and verbal responses and impaired memory and concentration. Apathy and withdrawal occur, but this must be distinguished from depression which occurs independently. Tremor and ataxia occur commonly. Progression leads to moderate or severe disability in over half of cases in three months, and acute exacerbations are common with intercurrent organic illness or psychotropic drugs. Focal signs may develop at any stage but the computer-assisted tomography (CT) scan shows only generalised cerebral atrophy in established cases (Fig. 8.8).

The next most common disorder of the central nervous system (CNS) is opportunistic infection with *Toxoplasma gondii* (Navia et al. 1986b, c). This usually presents

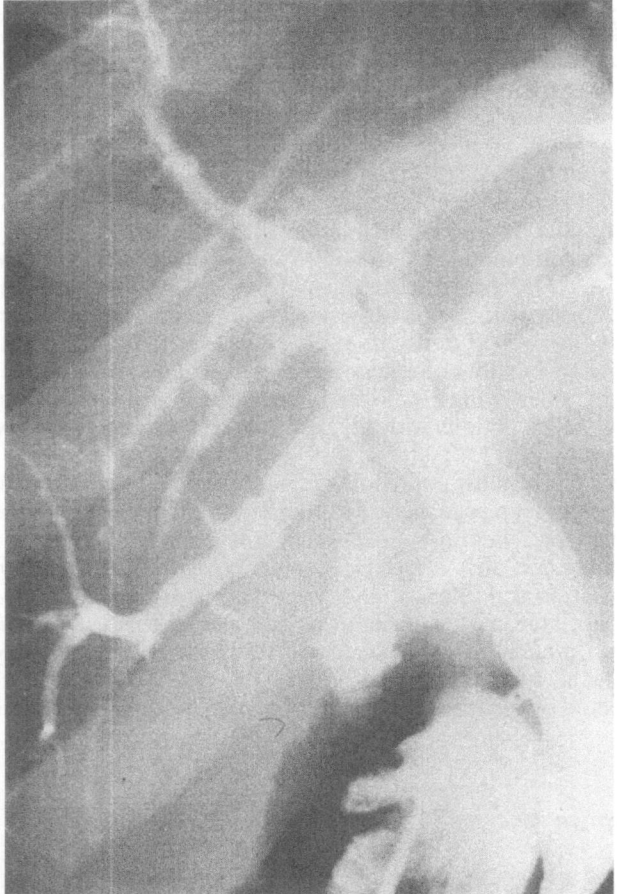

Fig. 8.7. Cholangitis: dilated common bile duct with stricture at lower end and irregularities of extra- and intrahepatic ducts (from Weller 1987).

with headache, fever and focal signs. The CT scan reveals hypodense lesions with ring contrast enhancement (Fig. 8.9). Double-dose contrast improves the rate of detection, but even then a few cases will be detected only on repeat scanning. The differential diagnosis includes primary CNS lymphoma, fungal, tuberculous or other bacterial abscess. In patients with meningism and a normal CT scan, the commonest finding is cryptococcal meningitis, diagnosed by both indian-ink staining of the cerebrospinal fluid and antigen detection. The spinal cord may be involved by a vascular myelopathy or a transverse myelitis that has been attributed to herpes zoster, or herpes simplex, CMV or HIV itself. Two forms of peripheral neuropathy may be encountered (a symmetrical sensorimotor type that may be painful, or a mononeuritis), in both of which HIV itself is implicated. CMV retinitis causes progressive visual failure, with a characteristic pattern of exudates and haemorrhages described as "cottage cheese and jam".

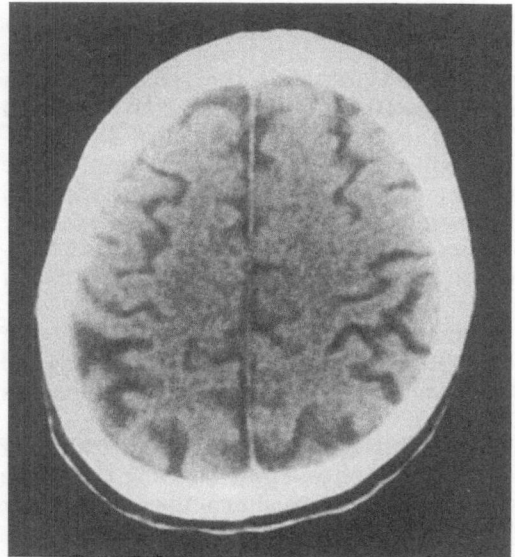

Fig. 8.8. CT scan: cerebral atrophy of HIV encephalopathy – widened sulci.

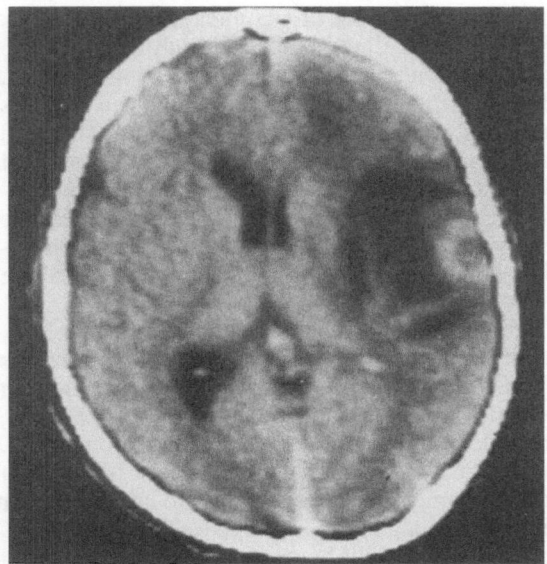

Fig. 8.9. CT scan: ring shadow of cerebral toxoplasmosis with surrounding oedema.

Tumours

Kaposi's Sarcoma. Prior to the epidemic of HIV, KS was a rare tumour in North America and Europe, with an annual incidence of 0.02 – 0.06 per 100 000. It occurred mainly in men over 50 years of age, of Jewish or Mediterranean ancestry. This classical form of the disease is usually confined to the lower limbs, runs an indolent course and responds well to radiotherapy or chemotherapy (Volberding et al. 1983; Safai and Weiss 1984). The tumour occurs in immunocompromised patients, especially renal allograft recipients, in whom it is more aggressive. Withdrawal of immunosuppression in this situation leads to tumour regression in as many as 50% (Harwood et al. 1979).

However, in Central Africa KS is a common tumour and accounts for some 9% of malignancies. The majority resembles the classical form seen in North America and Europe but a more aggressive disease with lymph node and visceral involvement has been described in children and young adults.

KS alone in the developed world is the second commonest presenting feature of AIDS (26% of reported cases) after PCP (51% of cases), 8% having both. The tumour is significantly more common amongst homosexual men than in the other groups at risk, who present more frequently with opportunistic infections. Its aetiology is unknown but this observation might suggest that a sexually transmitted agent other than HIV may be involved. CMV infection has been implicated in both the classical and HIV-related disease on the basis of seroepidemiological studies and the finding of virus particles and viral nucleic acid in tumour tissue. The role of CMV is unclear. Other workers have not confirmed these findings and, if present, the virus may only be a passenger. A genetic predisposition, with a higher frequency of HLA-DR5 in both classical and HIV-related disease, has also been described.

HIV-related KS is characterised by widespread skin, mucous membrane, visceral and lymph node involvement (Friedman-Kien et al. 1982). Skin lesions are the most common presenting complaint. They appear as pink or red macules or violaceous plaques and nodules on the face, trunk or limbs. Early skin lesions may be difficult to differentiate from other benign skin conditions such as granulomas, bruises, naevi, dermatofibromata, secondary syphilis or lichen planus. Histologically, the tumours consist of spindle-shaped cells arranged in broad bands, with vascular slits and extravasation of erythrocytes between the cells. The histological appearances of HIV-related and classical KS are indistinguishable. The median survival from diagnosis is about 30 months, with some patients running a more rapid, fulminant course and others remaining well for several years, with minimal cutaneous disease. Supervening opportunistic infection is the commonest problem in patients with KS alone and the usual terminal event.

The gastrointestinal tract is one of the commonest internal organs to be involved. If upper and lower gastrointestinal endoscopy is performed at presentation, lesions will be demonstrated in about 40% of patients. At post mortem they are present in more than 70%. Involvement of the hard palate and alveolar ridges, oropharynx, oesophagus, stomach, duodenum, colon and rectum have been demonstrated. Lesions resemble the range seen in the skin from small, flat telangiectatic lesions, not well demonstrated by contrast studies and only seen at endoscopy, to larger nodular or polypoid lesions. Endoscopic biopsy has a high false negative rate with only 23% of suspicious lesions being confirmed histologically because of their predominant position deep in the submucosa (Friedman et al. 1985).

Complications from involvement of the gut are unusual. Haemorrhage from lesions either acute or chronic leading to iron deficiency anaemia may occur. Several cases of KS presenting as an acute inflammatory-bowel-like syndrome, with diarrhoea and ulceration on barium enema, have also been described (Weber et al. 1985). A protein losing enteropathy may also occur.

Except for these circumstances, endoscopy is not routinely required to demonstrate visceral involvement if the diagnosis is made on the basis of mucocutaneous lesions. However, it is used in clinical trials to stage adequately the extent of the tumour. The staging system used for classical KS is inappropriate because the locally indolent and aggressive forms of tumour are rare and an alternative staging system has been suggested (Table 8.16).

Table 8.16. Staging of Kaposi's sarcoma

		Classical	HIV-related
Stage I	(A + B)[a]	Local indolent	Limited cutaneous (1 anatomical region)
Stage II	(A + B)	Local aggressive	Dissiminated cutaneous (>1 anatomical region)
Stage III	(A + B)	Generalised mucocutaneous/ lymphadenopathic	Visceral only (e.g. lymph node/GI tract)
Stage IV	(A + B)	Visceral	Cutaneous and visceral

[a]GI, gastrointestinal.
[a]A, no symptoms; B, non-specific symptoms, e.g. fever and weight loss.

Lymphoma. Early studies identified small numbers of homosexual men with non-Hodgkin's lymphoma developing in a setting of PGL, opportunistic infections and KS. According to the original definition of AIDS, homosexual men with lymphomas other than primary cerebral could not be notified as AIDS because lymphomas were considered to be a known cause of immunosuppression. However, Cancer Registry data in San Francisco and Los Angeles indicated up to a three-fold rise in high-grade lymphomas in young never-married men in 1983 and the characteristics of 90 such cases in homosexual men were reported (Ziegler et al. 1984). The tumours are of B cell origin and may present de novo, in a setting of prodromal lymphadenopathy, opportunistic infection or KS. The majority of patients in this series presented with extranodal involvement, predominantly in the central nervous system, bone marrow and gut (Table 8.17). The survival and response to treatment was poor. These features are similar to those of the generalised and aggressive KS of AIDS. Non-Hodgkin's lymphoma is now recognised as another manifestation of AIDS, but the mechanisms involved in the transition from the follicular hyperplasia of PGL with polyclonal B cell activation through to B cell lymphoma and the role of HIV and other human T cell trophic viruses and EBV have yet to be fully determined.

Management

The management of HIV infection begins with the antibody test itself. There is a temptation to treat this test like any other. However, it is recommended that no patient should have their HIV antibody status determined without their consent. Counselling should be provided to prepare a patient before the test and give support and advice when the result is communicated. Such counselling may be very time

Table 8.17. Extranodal sites of non-Hodgkin's lymphoma in 88 patients

CNS	
Brain mass	21
Other	24
Bone marrow	30
GI tract	22
Lung	8
Liver	8
Skin	7
Other	7

CNS, central nervous system.
Modified from Ziegler et al. 1984.

Table 8.18. Pre- and post-test counselling (checklist)

Pre-test	Post-test (positive)
Determine expectation of outcome	Response to result
Review what test does and does not reveal	Risk of developing AIDS
Avoiding further transmission/infection	Avoiding further transmission/infection
Who to tell?	Who to tell?
Problems re: insurance/mortgage	Sources of support
Leaflets	Follow-up

consuming (see Chap. 9). There are a few short cuts if one is to obtain the informed consent of the individual and many physicians, health advisers and counsellors use a checklist, so that important information is not forgotten (Table 8.18). It is more difficult to maintain confidentiality in a busy casualty or general out-patient setting than it is in a genito-urinary medicine clinic. Nevertheless we have an obligation to strive to maintain it and this should be recognised by all concerned. The consequences for an individual if confidentiality is breached may be devastating. There are many accounts of ostracism, loss of employment or accommodation and even suicide. The test has been used as a substitute for an adequate history. It is also used on the ground of infection control when a raising of routine standards that will prevent cross-infection would be more appropriate.

In the author's department, asymptomatic anti-HIV positive patients are seen three to four monthly. This interval between appointments is arbitrary, but in our experience longer intervals between check-ups in the asymptomatic patient are unsatisfactory. In cohort studies, the annual progression rate to AIDS is 6%–7%. With this information, HIV-infected patients demand and require constant reassurance that all is well as many live "waiting for the worst" (Miller and Green 1985). They need open access to immediate consultation if they are symptomatic. Symptoms of anxiety or depression, with fatigue, impairment of memory, poor concentration, shortness of breath, sweating, loss of appetite, weight loss and diarrhoea are often difficult for both physician and patient to differentiate from the more sinister life-threatening complications of HIV infection, with its major presentations affecting the nervous, respiratory and gastrointestinal systems.

At each appointment, a full history is taken; the patient is weighed and examined completely. Particular attention is paid to the skin, mouth, lymph nodes, respiratory system and anogenital areas for signs of chronic HIV infection and, more importantly, early signs of an opportunistic infection or tumour. A full blood count, differential white cell count and erythrocyte sedimentation rate, together with the history and examination, will provide the physician with considerable information as to the severity of chronic HIV infection (Carne et al. 1987a).

These appointments in conjunction with a counsellor also serve as a useful opportunity to repeat discussions on safer sex guidelines (see Table 8.10). A decrease in the rate of rise of HIV infection amongst homosexual men, a continuing decrease in the prevalence of gonorrhoea and considerable changes in sexual behaviour have recently been documented in our department (Carne et al. 1987b). Furthermore, anxieties and fears often precipitated by misinformation from friends or the media can be relieved, and social problems such as prejudice in the work place and housing difficulties attended to .

The therapy of HIV infection has been limited largely to the treatment of its malignant and infectious complications. Cytotoxic therapy for the tumours may lead to further immunosuppression and the risk of life threatening opportunistic infection. Most of the infections are due to reactivation of latent organisms in the host or in some cases to ubiquitous organisms to which we are continuously exposed. The treatment of these infections tends to suppress rather than eradicate organisms, so relapse is common when therapy is stopped. Furthermore the side-effects of many of the drugs used do not easily facilitate the long-term therapy that is required.

Kaposi's Sarcoma

The median survival from time of diagnosis, for patients with KS alone, is substantially better than for those with opportunistic infection (31 versus 9 months). In the absence of any evidence for improved survival with early therapy, the first lesions of KS may require no treatment. Local radiotherapy can be used for treatment of mucocutaneous lesions for cosmetic reasons and for larger lesions causing local complications, such as those in the oropharynx. Systemic chemotherapy is used in generalised mucocutaneous disease with or without visceral involvement. Various single-agent and combined cytotoxic regimes have been used with their many side-effects, risk of further immunosuppression and reactivation of latent infection. Lymphoblastoid and recombinant human alpha-interferons have been used with a few complete remissions. Some 30%–40% of patients have partial remissions with this treatment, but severe side-effects are common, with constitutional symptoms and cytopaenias. At present the best that we should expect with any form of systemic therapy is a temporary, partial remission at considerable cost to the patient in terms of adverse effects and with no evidence yet that the treatment will alter the long-term prognosis. Many centres therefore have a conservative policy for the use of such therapy in most cases of KS.

Viral Infections (Table 8.19)

Herpes simplex infections respond to oral acyclovir (Mindel et al. 1984). Prophylactic treatment may be required as severe, frequent recurrences are common in chronic HIV infection short of AIDS. Some patients respond to less frequent doses. Intravenous acyclovir is preferred for severe persistent mucocutaneous infections.

Intravenous acyclovir has been used in other immunocompromised patients for the treatment of herpes zoster infection to prevent dissemination (Shepp et al. 1986).

Table 8.19. Therapy of viral infections

Infection	Drug	Dose	Duration	Route	Side-effects
Herpes simplex	Acyclovir	5–10 mg/kg, 8 hourly	10–14 days	i.v.	
	Acyclovir	200 mg, 5 times daily	10–14 days	Oral	Minimal
	Prophylaxis	200 mg q.i.d. (Lower frequency possible)	?Indefinite	Oral	Minimal
Varicella zoster	Acyclovir	10 mg/kg, 8 hourly	10 days	i.v.	Minimal
	Acyclovir	400 mg q.i.d. 800 mg q.i.d. being evaluated	10 days	Oral	Minimal
Cytomegalo-virus	Ganciclovir	2.5–5 mg/kg, 8 hourly	14–21 days	i.v.	Marrow suppression
	Maintenance	2.5–5 mg/kg, probably daily	Indefinite	i.v.	Marrow suppression
	Phosphono-formate	0.05–0.16 mg/kg per min	14–21 days	i.v. constant infusion	Renal impairment, bone accumulation

i.v., intravenous; q.i.d., four times daily.

Dissemination appears to be unusual in HIV infection, but it may be given orally at a dose of 400 mg four times, daily.

Ganciclovir (DHPG, 9-(1,3-dihydroxy-2-propoxymethyl)guanine), like acyclovir is an acyclic analogue of deoxyguanosine. It has shown promising results in uncontrolled studies in the treatment of CMV retinitis and to a lesser extent colitis (Collaborative DHPG Treatment Study Group 1986). Patients with pneumonitis and encephalitis do not respond as well. In the retinopathy it would appear to delay the progression of disease, but maintenance therapy is required and even then relapse or progression may occur. Phosphonoformate (Foscarnet, a pyrophosphate analogue) has also been used in CMV infections with some success (Weber et al. 1987b). It inhibits polymerase enzymes, but it has to be given as a continuous intravenous infusion. Intravenous therapy with both agents is a major drawback to antiviral maintenance therapy but many patients are being maintained without complications at home, using Hickman catheters. Hairy leucoplakia responds to oral acyclovir.

Protozoal Infections (Table 8.20)

Of the protozoal infections, pneumocystis is the most likely to respond well. Cotrimoxazole remains the treatment of choice. High-dose intravenous therapy for three weeks is the standard regime, although once the patient's fever has settled and the shortness of breath and blood gases have improved the drug can be given orally. A minor rash during treatment may be ignored, but a very itchy generalised eruption with recurrence of fever, or severe cytopenia may occur at 7–10 days and requires a change to pentamidine mesylate. This needs to be given by slow intravenous infusion to avoid the persistent pain and sterile abscesses that occur with intramuscular injection. Hypoglycaemia and renal failure may occur during treatment. Large, controlled studies of prophylaxis have not been conducted following first attacks of PCP in AIDS. However, many physicians use daily low-dose cotrimoxazole two to four tablets daily, or one tablet of pyrimethamine-sulphadoxine weekly (Fansidar), but with both these sulphonamide sensitivity may be a problem. Monthly intramuscular pentamidine has also been used. There have been several recent advances in the manage-

Table 8.20. Therapy of protozoal infections

Infection	Drug	Dose	Duration	Route	Side-effects
Pneumocystis carnii pneumonia	Cotrimoxazole	20 mg/kg per day of trimethoprim	14–21 days	i.v. then oral	Nausea, fever, rash Marrow suppression
	Pentamidine isethionate	4 mg/kg per day	14–21 days	i.v. as slow single daily	Hypotension, hypo- glycaemia, renal
	Pentamidine mesylate	2.5 mg/kg per day day	14–21 days	infusion	failure, hepatitis, marrow suppression
Toxoplasmosis	Sulphadiazine	2–4 g daily	Indefinite	Oral	As for sulphonamides
	Pyrimethamine	25 mg daily	Indefinite	Oral	(above)
	Clindamycin	500 mg q.i.d.	Indefinite	Oral	
Isosporiasis	Cotrimoxazole	2 tabs q.i.d.	Indefinite	Oral	As for sulphonamides (above)

i.v., intravenous; q.i.d., four times daily.

ment of PCP. Once patients develop respiratory failure and require assisted mechan- ical ventilation, mortality is high. Limited studies have demonstrated a beneficial effect of a short course of high-dose corticosteroids in this situation (MacFadden et al. 1987). Nebulised pentamidine is also showing great promise in the acute phase. This reduced systemic toxicity is an advantage and in particular will be useful in patients on concurrent Zidovudine (3'-azido-3'-deoxythymidine, formerly azido- thymidine (AZT) or DHPG therapy is that further bone marrow depression would produce prohibitive toxicity. Trimetrexate is another therapy showing promise.

Cryptosporidiosis may repond to spiramycin (1 g four times, daily) (Portnoy et al. 1984) or a combination of quinine and clindamycin, but reported success is anec- dotal. Symptoms and excretion of cysts may be intermittent and so spontaneous remission may occur. There are also anecdotal reports of a response to interleukin-2 (Kern et al. 1985). Symptomatic treatment with codeine phosphate, loperamide and other drugs may be the only effective measure. Isosporiasis responds well to co- trimoxazole, but relapses occur in 50% of cases with cessation of treatment (De- Hovitz et al. 1986).

Cerebral toxoplasmosis responds well if treatment is instituted early and a combi- nation of sulphadiazine and pyrimethamine is the treatment of first choice. Side- effects may require stopping the sulphadiazine but clindamycin has been used suc- cessfully as a substitute in uncontrolled studies. Dexamethasone is occasionally used in severe cases, in a short course, to reduce cerebral oedema. Relapse is common following cessation of therapy and treatment should continue indefinitely.

Bacterial Infections

Bacterial pneumonias will respond to conventional antibiotics. Bacterial skin infec- tions also often require systemic antibiotics but prophylaxis with Hibiscrub soap is often useful.

Salmonella species infections respond to treatment with appropriate antibiotics, but relapses of enteritis and/or bacteraemia are common.

Diarrhoea may respond to metronidazole, even in the absence of recognised pathogens in the stool. Symptomatic treatment with codeine phosphate, loperamide and other drugs can be used, and may be the only effective treatment for crypto- sporidiosis. *Mycobacterium tuberculosis* is treated conventionally. The atypical organisms are resistant to conventional anti-tuberculous therapy. Ansamycin (a rifamycin derivative) and clofazimine (an anti-leprosy compound) are among the agents being tried, having been shown to have some in-vitro activity, but the treatment of patients with disseminated infection has been largely unsuccessful. Until a specific therapy is found, the exact pathogenic role of these organisms in the many clinical symptoms of AIDS will be difficult to elucidate.

Fungal Infections

Seborrhoeic dermatitis associated with *Pityrosporum* infection responds to antifungal and/or steroid creams. Other dermatophytic fungal infections respond to imidazole creams. Oral *Candida* is often asymptomatic in its early stages and may not require therapy. In more severe infections, local treatment with frequent nystatin suspension or pastilles or amphotericin lozenges can be used. Systemic therapy with ketoconazole 200–400 mg orally daily may be required, and is the drug of choice for oesophageal candidiasis. Long-term therapy may be required to prevent recurrences and liver function tests should be monitored. Cryptococcal and other systemic fungal infections require treatment with amphotericin with or without flucytosine for a minimum of six weeks. Fluconazole is a new systemic antifungal that is being evaluated. Its advantages are low toxicity (in particular hepatic), oral administration and good penetration of the cerebrospinal fluid.

Antivirals

The ideal antiviral agent should be specific, orally absorbed and cross the blood/brain barrier. It should also be free from adverse effects, since the best we can anticipate is suppression of productive viral replication with the problem of latently infected cells remaining. Theoretically, inhibition of productive viral replication may allow some recovery of immune function, perhaps encouraging regression of tumours and elimination of the conditions favouring opportunistic infections.

A variety of potential targets for antiviral therapy have now been identified from a better understanding of the replicative cycle and molecular biology of HIV. Most efforts are being focussed on specific inhibitors of the HIV reverse transcriptase enzyme.

Limited, uncontrolled, studies with Suramin and an antimony compound, HPA 23, showed little benefit and the adverse effects were unacceptable. Foscarnet has been shown in vitro to inhibit reverse transcriptase and limited controlled studies have so far failed to demonstrate a convincing clinical benefit.

The most promising group of reverse transcriptase inhibitors are the 2′,3′-dideoxynucleoside analogues. Zidovudine has been shown to have considerable activity against HIV in vitro. It is a competitive inhibitor of reverse transcriptase and a DNA chain terminator in vitro.

Zidovudine is orally absorbed and cerebrospinal fluid levels are approximately 50% of corresponding plasma levels. In a double-blind randomised controlled clinical trial of oral therapy (250 mg orally, 4 hourly for six months) in 280 patients with past PCP or severe symptomatic chronic HIV infection (AIDS-related complex), it has been shown significantly to reduce mortality and morbidity and to decrease episodes of opportunistic infection (Fischl et al. 1987; Richman et al. 1987). There are significant improvements in some immunological parameters and a significant antiviral action was demonstrated by a decrease in serum P24 (viral core protein). Limited uncontrolled studies have shown improvement in neurological disease. Side-effects include a megaloblastic anaemia, neutropenia, nausea, insomnia and myalgia. Dideoxycytidine is undergoing phase 1 studies, and preliminary data suggest that it is a more potent inhibitor of HIV on a molar basis and is less toxic.

Controlled trials of Zidovudine and other antiviral agents alone or in combination to reduce toxicity and improve efficacy are planned, in asymptomatic patients with laboratory markers that predict more rapid progression to AIDS. Because of the demonstrated efficacy of Zidovudine in patients with AIDS-related complex, this will be used as an end point in such trials. If found to be beneficial in this situation and provided that side-effects are minimal, controlled trials will then be carried out very early in HIV infection, even at the point of seroconversion.

Future aims will be to combine not only antiviral agents but also antiviral and immunomodulatory therapies (if effective and beneficial examples of the latter are found) in larger placebo controlled trials. Enhancement of the cellular immune response to HIV may be one way of attacking the virus in some of the sites where it persists.

Dr. I. V. D. Weller is a Wellcome Trust Senior Lecturer in Infectious Diseases.

References

Abrams DI, Moss T, Volberding PA (1985) Lymphadenopathy: end point prodrome? Update of a 36 month prospective study. Adv Exp Med Biol 187: 73–84

Angarano G, Pastore G, Monno L, Santantonio T, Luchena N, Schiraldi O (1985) Rapid spread of HTLV III infection among drug addicts in Italy. Lancet ii: 1302

Anonymous (1986) Who will get AIDS. Lancet ii: 953–954

Balthazar EJ, Megibow AJ, Fazzini E, Opulencia JF, Engel I (1985) Cytomegalovirus colitis in AIDS: radiographic findings in 11 patients. Radiology 155: 585–589

Barre-Sinoussi F, Cherman JC, Rey F et al. (1983) Isolation of T lymphotropic retrovirus for a patient at risk of acquired immune deficiency syndrome (AIDS). Science 220: 868–870

Barton SE, Underhill GS, Gilchrist C, Jeffries DJ, Harris JRW (1985) HTLV III antibody in prostitutes. Lancet ii: 1424

Biberfeld G, Brown F, Esparza J et al. (1987) WHO Working Group on characterization of HIV-related retroviruses: criteria for characterization and proposal for a nomenclature system. AIDS 1: 189–190

Brucker G, Brun-Vezinet F, Rosenheim M, Rey MA, Katlama C, Gentilini M (1987) HIV-2 infection in two homosexual men in France. Lancet i: 223

Carne CA, Tedder RS, Smith A et al. (1985a) Acute encephalopathy coincident with seroconversion for anti-HTLV III. Lancet ii: 1206–1208

Carne CA, Weller IVD, Sutherland S et al. (1985b) Rising prevalence of human T lymphotropic virus type III (HTLV III) infection in homosexual men in London. Lancet i: 1261–1262

Carne CA, Weller IVD, Loveday C, Adler MW (1987a) From persistent generalised lymphadenopathy to AIDS: who will progress? Br Med J 294: 868–869

Carne CA, Weller IVD, Johnson AM et al. (1987b) Prevalence of antibodies to human immunodeficiency virus, gonorrhoea rates and changed sexual behaviour in homosexual men in London. Lancet i: 656–658

Centers for Disease Control (1982) Update on acquired immune deficiency syndrome (AIDS) – United States. MMWR 31: 507–514

Centers for Disease Control (1983) Immunodeficiency among female sexual partners of males with acquired immune deficiency syndrome (AIDS) New York. MMWR 31: 697–698

Centers for Disease Control (1984) Update: acquired immunodeficiency syndrome (AIDS) United States. MMWR 33: 661–664

Centers for Disease Control (1985a) Revision of case definition of acquired immunodeficiency syndrome for national reporting – United States. MMWR 34: 373–375

Centers for Disease Control (1985b) Self-reported behavioral change among gay and bisexual men – San Francisco. MMWR 34: 613–615

Centers for Disease Control (1985c) Changes in premature mortality. MMWR 34: 669–671

Centers for Disease Control (1986a) Tuberculosis – United States 1985 and the possible impact of HTLV-III/LAV infection. MMWR 35: 74–76

Centers for Disease Control (1986b) Classification system for human T-lymphotropic virus type III/lymphadenopathy-associated virus infections. MMWR 35: 334–339

Centers for Disease Control (1986c) Update: acquired immunodeficiency syndrome – United States. MMWR 35: 757–766

Centers for Disease Control (1986d) Update: acquired immunodeficiency syndrome – United States. MMWR 36: 17–21

Centers for Disease Control (1987) Revision of the CDSC surveillance case definition for acquired immunodeficiency syndrome. MMWR 36 (suppl)

Chamberland ME, Castro KG, Haverkos HW et al. (1984) Acquired immunodeficiency syndrome in the United States: an analysis of cases outside high incidence groups. Ann Intern Med 101: 617–623

Clavel F, Guetard D, Brun-Vezinet F et al. (1986) Isolation of a new human retrovirus from West African patients with AIDS. Science 233: 343–346

Clumeck N, Van de Perre P, Carael M et al. (1985) Heterosexual promiscuity among African patients with AIDS. N Engl J Med 313: 182

Cockerill FR, Hurley DV, Malagelada JR et al. (1986) Polymicrobial cholangitis and Kaposi's sarcoma in blood product transfusion-related acquired immune deficiency syndrome. Am J Med 80: 1237–1241

Collaborative DHPG Treatment Study Group (1986) Treatment of serious cytomegalovirus infections with 9-(1,3-dihydroxy-2-propoxymethyl) guanine in patients with AIDS and other immunodeficiencies. N Engl J Med 314:801–805

Cooper DA, Gold J, Maclean P et al. (1985) Acute AIDS retrovirus infection. Lancet i: 537–540

Curran JMC, Morgan WM, Hardy AM, Jaffe HW, Darrow WW, Dowdle WR (1985) The epidemiology of AIDS: current status and future prospects. Science 229: 1352–1357

DeHovitz JA, Pape JW, Boncy M, Johnson WD (1986) Clinical manifestations and therapy of *Isospora belli* infection in patients with the acquired immunodeficiency syndrome. N Engl J Med 315: 87–90

Denning DW, Anderson J, Rudge P, Smith H (1987) Acute myelopathy associated with primary infection with human immunodeficiency virus. Br Med J 294: 143–144

Dobbins WV III, Weinstein WM (1985) Electron microscopy of the intestine and rectum in the acquired immunodeficiency syndrome. Gastroenterology 88: 738–749

Fischl MA, Rickman DD, Grieco MH (1987) The efficacy of Azidothymidine (AZT) in the treatment of patients with AIDS and AIDS-related complex. N Engl J Med 317: 185–191

Frager DH, Frager JD, Wolf EL et al. (1986) Cytomegalovirus colitis in acquired immune deficiency syndrome: radiologic spectrum. Gastrointest Radiol 11: 241–246

Friedman SL, Wright TL, Atlman DF (1985) Gastrointestinal Kaposi's sarcoma in patients with acquired immunodeficiency syndrome. Endoscopic and autopsy findings. Gastroenterology 89: 102–108

Friedman-Kien AE (1986) Viral origin of hairy leucoplakia. Lancet ii: 694

Friedman-Kien A, Laubenstein L, Marmor M et al. (1981) Kaposi's sarcoma and pneumocystis among homosexual men – New York City and California. MMWR 30: 305–308

Friedman-Kien A, Laubenstein LJ, Rubinstein P. et al. (1982) Disseminated Kaposi's sarcoma in homosexual men. Ann Intern Med 96: 693–700

Gallo RC, Savin PS, Gelman EP et al. (1983) Isolation of human T cell leukaemia virus in acquired immune deficiency syndrome (AIDS). Science 220: 865–867

Gartner S, Markovitz P, Markovitz DM, Kaplan MH, Gallo RC, Popovic M (1986) The role of mononuclear phagocytes in HTLV III/LAV infection. Science 233: 215–219

Goedert JJ, Biggar RJ, Weiss SH et al. (1986) Three year incidence of AIDS in five cohorts of HTLV III infected risk group members. Science 231: 992–995

Gordon SC, Reddy KR, Gould EE et al. (1986) The spectrum of liver disease in the acquired immunodeficiency syndrome. J Hepatol 2: 475–484

Gottlieb MS, Schanker HM, Fan PT, Saxon A, Weisman JD, Pzalski I (1981a). *Pneumocystis pneumonia* – Los Angeles. MMWR 30: 250–252

Gottlieb MS, Schroff R, Schanker HM et al. (1981b) *Pneumocystis carinii* pneumonia and mucosal candidiasis in previously healthy homosexual men: evidence of a new acquired cellular immunodeficiency. N Engl J Med 305: 1425–1431

Greenspan JS, Greenspan D, Lennette ET et al. (1985) Replication of Epstein–Barr virus within the epithelial cells of oral "hairy" leukoplakia, an AIDS-associated lesion. N Engl J Med 313: 1564–1571

Grizburg HM, Weiss SH, MacDonald J et al. (1988) HTLV III exposure among drug users. Cancer Res (in press)

Gross TL, Wheat J, Bartlett M, O'Connor KW (1986) AIDS and multiple system involvement with *Cryptosporidium*. Am J Gastroenterol 81: 456–458

Hardy AM, Rausch KJ, Curran JW (1985) The economic impact of the first 10,000 cases of AIDS in the United States. JAMA 255: 209–211

Harris C, Small CB, Klein RJ et al. (1983) Immunodeficiency in female sexual partners of men with the acquired immunodeficiency syndrome. N Engl J Med 308: 1181–1184

Harwood AR, Osoba D, Hofstader SL (1979) Kaposi's sarcoma in recipients of renal transplants. Am J Med 67: 759–765

Hymes KB, Cheung T, Greene JB (1981) Kaposi's sarcoma in homosexual men – a report of eight cases. Lancet ii: 598–600

Jarrett DRJ, Zeegen R (1986) Recurrent typhoid in an HTLV-III antibody positive man. Gut 27: 587–588

Jesson WJ, Thorp RW, Mortimer PP, Oates JK (1985) Prevalence of anti-HTLV III in UK risk groups 1984/5. Lancet i: 155

Johnson AM, Adler MW, Crown JM (1986) The acquired immune deficiency syndrome and epidemic of infection with human immunodeficiency virus: costs of care and prevention in an Inner London District. Br Med J 293: 489–492

Judson FN (1983) Fear of AIDS and gonorrhoea rates in homosexual men. Lancet ii: 59–60

Kavin H, Jonas RB, Chowdhury L, Kabin S (1986) Acalculous cholecystitis and cytomegalovirus infection in the acquired immunodeficiency syndrome. Ann Intern Med 104: 53–54

Kern P, Toy J, Dietrich M (1985) Preliminary clinical observations with recombinant Interleukin-2 in patients with AIDS or LAS. Blut 50: 1–6

Kotler DP, Gaetz HP, Lange M, Klein EB, Holt PR (1984) Enteropathy associated with the acquired immunodeficiency syndrome. Ann Intern Med 101: 421–428

Kreiss JK, Kitchen LW, Prince HE et al. (1985) Antibody to human T lymphotropic virus type III in wives of haemophiliacs. Evidence for heterosexual transmission. Ann Intern Med 102: 623–626

Lange JMA, Paul DA, Huisman HG et al. (1986) Persistent HIV antigenaemia and decline of HIV core antibodies associated with transition to AIDS. Br Med J 293: 1459–1462

Lange JMA, de Wolf F, Krone WJA, Danner SA, Coutinho RA, Goudsmit J (1987) Decline of antibody reactivity to outer viral core protein P17 is an earlier serological marker of disease progression in human immunodeficiency virus infection than anti-P24 decline. AIDS 1: 155–165

Lawrence J (1985) The immune system in AIDS. Sci Am 253: 70–79

Ma P, Soave R (1983) Three-step stool examination for cryptosporidiosis in ten homosexual men with protected watery diarrhoea. J Infect Dis 147: 824–828

MacFadden DK, Edelson JD, Hyland RH, Rodriguez CH, Nouye T, Rebuck AS (1987) Corticosteroids as adjunctive therapy in treatment of *Pneumocystis carinii* pneumonia in patients with acquired immunodeficiency syndrome. Lancet i: 1477–1479

Macher AM (1984) Infection in the acquired immunodeficiency syndrome. In: Acquired immunodeficiency syndrome: epidemiologic, clinical, immunologic and therapeutic considerations (Moderator Fauci AS) pp 94–96. Ann Intern Med 100: 92–100

Marasca G, McEvoy M (1986) Length of survival of patients with acquired immune deficiency syndrome in United Kingdom. Br Med J 292: 1727–1729

Margulis SJ, Honig CL, Soave R, Govoni AF, Monradian JA, Jacobson IM (1986) Biliary tract obstruction in the acquired immunodeficiency syndrome. Ann Intern Med 105: 207–210

Mathur-Wagh M, Enlow RW, Sprigland I et al. (1984) Longitudinal study of persistent generalised lymphadenopathy in homosexual men: relation to the acquired immunodeficiency syndrome. Lancet i: 1033–1038

Mathur-Wagh M, Mildvan D, Serice RT (1985) Follow up at 4½ years on homosexual men with generalised lymphadenopathy. N Engl J Med 313: 1542–1543

Metroka CE, Cunningham-Rundles S, Pollack MS et al. (1983) Generalised lymphadenopathy in homosexual men. Ann Intern Med 99: 585–591

Miller D, Green J (1985) Psychological support and counselling for patients with acquired immune deficiency syndrome (AIDS) Genitourin Med 61: 273–278

Mindel A, Weller IVD, Faherty A et al. (1984) Prophylactic oral acyclovir in recurrent genital herpes. Lancet ii: 57–59

Modigliani R, Bories C, Le Charpentier Y et al. (1985) Diarrhoea and malabsorption in acquired immune deficiency syndrome: a study of four cases with special emphasis on opportunistic protozoan infestations. Gut 26: 179–187

Mortimer PP, Jesson WJ, Vandervelde EM, Pererra MS (1985a) Prevalence of antibody to human T lymphotropic virus type III by risk group and area. United Kingdom 1978–84. Br Med J 290: 1176–1178

Mortimer PP, Vandervelde EM, Jesson WJ, Pererra MS (1985b) HTLV III antibody in Swiss and English intravenous drug abusers. Lancet ii: 449–450

Murphy MF, Metcalfe P, Waters AH et al. (1987) Incidence and mechanism of neutropenia and thrombocytopenia in patients with human immunodeficiency virus infection. Br J Haematol 66: 337–340

Murray JF, Felton CP, Garay SM et al. (1984) Pulmonary complications of the acquired immunodeficiency syndrome. N Engl J Med 310: 1682–1688

Nadelman RB, Mathur-Wagh U, Vancovitz SR, Mildvan D (1985) Salmonella bacteremia associated with the acquired immunodeficiency syndrome (AIDS). Arch Intern Med 145: 1968–1971

Navia BA, Petito CK, Gold JWM, Cho E-S, Jordan BD, Price RW (1986a) Cerebral toxoplasmosis complicating the acquired immune deficiency syndrome: clinical and neuropathological findings in 27 patients. Ann Neurol 19: 224–238

Navia BA, Jordan BD, Price RW (1986b) The AIDS dementia complex: I. Clinical features. Ann Neurol 19: 517–524

Navia BA, Cho E-S, Petito CK, Price RW (1986c) The AIDS dementia complex: II. Neuropathology. Ann Neurol 19: 525–535

Orenstein MS, Tavitian A, Yonk B et al. (1985) Granulomatous involvement of the liver in patients with AIDS. Gut 26: 1220–1225

Piot P, Quinn TC, Taelman H et al. (1984) Acquired immunodeficiency syndrome in a heterosexual population in Zaire. Lancet ii: 65–69

Polk BF, Fox R, Brookmeyer R et al. (1987) Predictors of the acquired immunodeficiency syndrome developing in a cohort of seropositive homosexual men. N Engl J Med 316: 61–66

Polsky B, Gold JWM, Whiimbey E et al. (1986) Bacterial pneumonia in patients with the acquired immunodeficiency syndrome. Ann Intern Med 104: 38–41

Portnoy D, Whiteside ME, Buckley E III, MacLeod CL (1984) Treatment of intestinal cryptosporidiosis with Spiramycin. Ann Intern Med 101: 202–204

Price RW, Navia BA, Cho E-S (1986) AIDS encephalopathy. Neurol Clin 4: 285–301

Rashleigh-Belcher HJC, Carne CA, Weller IVD, Smith AM, Russell RGG (1986) Surgical biopsy for persistent generalised lymphadenopathy. Br J Surg 73: 183–185

Redfield RR, Markham PD, Salahuddin SZ et al. (1985) Frequent transmission of HTLV III among spouses of patients with AIDS-related complex and AIDS. JAMA 253: 1571–1573

Richman DD, Fischl MA, Grieco MH et al. (1987) The toxicity of Azidothymidine (AZT) in the treatment of patients with AIDS and AIDS-related complex. N Engl J Med 317: 192–197

Rivin BE, Monroe JM, Hubschman BA et al. (1984) AIDS outcome: a first follow up. N Engl J Med 311: 857

Rotterdam H, Sommers SC (1985) Alimentary tract biopsy lesions in the acquired immune deficiency syndrome. Pathology 17: 181–192

Safai B, Weiss H (1984) Clinical manifestations of Kaposi's sarcoma. In: Ma P, Armstrong D (ed) AIDS and infections of homosexual men. Yorke Medical Books, New York, Chap 16, pp 211–214

Sarngadharan MG, Popovic M, Bruch L et al. (1984) Antibodies reactive with human T lymphotropic retrovirus (HTLV III) in the serum of patients with AIDS. Science 224: 506–508

Shepp DH, Dandliker PS, Meyers JD (1986) Treatment of varicella-zoster virus infection in severely immunocompromised patients. A randomised comparison of acyclovir and vidarabine. N Engl J Med 314: 208–212

Skitovsky AA, Cline M, Lee PR (1986) Medical costs of patients with AIDS in San Francisco. JAMA 256: 3103–3106

Soave R, Danner RL, Honig CL, Ma P, Hart CC, Nash T, Roberts RB (1984) Cryptosporidiosis in homosexual men. Ann Intern Med 100: 504–511

Stricker RB, Abrams DI, Corash L, Shuman MA (1985) Target platelet antigen in homosexual men with immune thrombocytopenia. N Engl J Med 313: 1375–1380

Van de Perre P, Rouvroy D, LePage P et al. (1984) Acquired immunodeficiency syndrome in Rwanda. Lancet ii: 62–65

Van de Perre P, Clumeck N, Carael M et al. (1985) Female prostitutes: a risk group for infection with human T lymphotropic virus type III. Lancet ii: 524–527

Volberding P, Conant MA, Stricker RB, Lewis BJ (1983) Chemotherapy in advanced Kaposi's sarcoma. Am J Med 74: 652–656

Walsh C, Krigel R, Lennette E, Karpatkin S (1985) Thrombocytopenia in homosexual patients. Ann Intern Med 1985; 103: 542–545

Weber JN, Carmichael DJ, Boylston A, Munro A, Whitear WP, Pinching AJ (1985) Kaposi's sarcoma of the bowel – presenting as apparent ulcerative colitis. Gut 26: 295–300

Weber JN, Clapham PR, Weiss RA et al. (1987a) Human immunodeficiency virus infection in two cohorts of homosexual men: neutralising sera and association of anti-gag antibody with prognosis. Lancet i: 119–122

Weber JN, Thom S, Barrison I (1987b) Cytomegalovirus colitis and oesophageal ulceration in the context of AIDS: clinical manifestations and preliminary report of treatment with Foscarnet (phosphonoformate). Gut 28: 482–487

Weiss RA, Clapham PR, Cheingsong-Popov R, Dalgleish AG, Carne CA, Weller IVD, Tedder RS (1985) Neutralising antibodies to human T-lymphotropic virus type III. Nature 316: 69–71

Weiss SJ, Goedert JJ, Sarngadharan MG et al. (1985) Screening test for HTLV III (AIDS agent) antibodies specificity, sensitivity and applications. JAMA 253: 221–225

Weller IVD (1987) Gastrointestinal and hepatic manifestations of AIDS. In: Adler MW (ed) British Medical Journal ABC of AIDS. British Medical Journal, London.

Weller IVD, Hindley DJ, Adler MW, Meldrum JT (1984) Gonorrhoea in homosexual men and media coverage of the acquired immune deficiency syndrome in London 1982–83. Br Med J 289: 1041

Wong-Staal F, Gallo RC (1985) Human T-lymphotrophic retroviruses. Nature 317: 395–403

Ziegler JL, Drew WI, Miner RC et al. (1982) Outbreak of Burkitt's like lymphoma in homosexual men. Lancet ii: 631–633

Ziegler JL, Beckstead JH, Voldberding P et al. (1984) Non-Hodgkin's lymphoma in 90 homosexual men: relation to generalised lymphadenopathy and the acquired immunodeficiency syndrome. N Engl J Med 311: 565–570

Chapter 9

AIDS: Counselling and Support

1. Hospital and Statutory Services
L. Glover and D. Miller

Introduction

The acquired immune deficiency syndrome (AIDS) is the most preventable of diseases but prevention can only be effective if people know with what they are dealing and what behavioural changes are required. Counselling for human immunodeficiency virus (HIV) and AIDS involves facilitating an understanding of the infection and of ways in which its spread may be avoided. The significance of the role of counselling can best be appreciated if one recognises that AIDS is not simply a medical problem with some social issues attached; rather it is a social problem with serious medical issues attached (Miller 1987a). Accordingly, effective counselling requires that we address a broad range of social concerns in addition to known facts and hypotheses about the transmission of infection through various social subgroups and activities. These issues must then be distilled into easily accessible and relevant health education. If one is to consider the role of the counsellor in its broadest sense, it includes a wide range of responsibilities and commitments. Clearly the HIV and AIDS counsellor has an extremely important role in supporting patients, their loved ones, families and carers, but in addition he or she is in an ideal position to provide health education not only to those requesting an HIV test and to those who are antibody positive, but also in the community, for example in schools and workplaces.

In order to pursue their broad responsibilities effectively, counsellors are required to keep up to date with rapidly changing medical information and a detailed knowledge of patient life-styles; they must be able to identify and perhaps to manage psychological, psychiatric and neurological issues and to tease apart behavioural phenomena from organic phenomena.

Family and marital therapy skills are often needed. The care of AIDS patients in a hospital setting should be based on a *team-centred* approach in order to ensure consistency and to avoid confusion (Christ and Wiener 1985; Wolcott et al. 1985; Miller and Green 1986). Counsellors must be able to work as part of that team and must liaise closely with medical and nursing staff. All patients must, of course, be managed in the context of strict confidentiality.

Finally, in order for staff to fulfil a counselling role successfully, adequate support and supervision is of paramount importance (McKusick 1986). This chapter will address these concerns and provide details of a successful counselling structure that has been in operation in the UK for over four years.

Psychological/Psychiatric Phenomena

Results from studies in many parts of the world uniformly reveal that knowledge of personal seropositivity or HIV disease evokes a wide range of possible severe psychological morbidity (Christ and Wiener 1985; Dilley et al. 1985; Miller and Green 1986; Wolcott et al. 1985). These studies and many others describe such affected persons as experiencing high levels of chronic anxiety, depression, and obsessional disturbance relating to issues described in Table 9.1.

Table 9.1. Psychological issues in HIV/AIDS counselling

Shock
Of diagnosis and possible death
Over loss of hopes for good news

Fear and anxiety
Over uncertain prognosis and course of illness
Of disfigurement and disability
Of effects of medication and treatment
Of isolation and abandonment and social/sexual rejection
Of infecting others and being infected by them
Of lover's ability to cope and their possible illness
Of loss of cognitive, physical, social and work abilities

Depression
Over "inevitability" of physical decline
Over absence of a cure
Over the virus controlling future life
Over limits imposed by ill-health and possible social, occupational,
   emotional and sexual rejection
From self-blame and recrimination for being vulnerable
   to infection in the first place

Anger and frustration
Over inability to overcome the virus
Over new and involuntary health/life-style restrictions
At being "caught out" and the uncertainty of the future

Guilt
Over past "misdemeanours" resulting in illness "punishment"
Over possibly having spread infection to others
Over being homosexual or a drug user

Obsessive disorders
Relentless searching for new diagnostic evidence on body
Faddism over health and diets
Preoccupations with death and decline, and avoidance of new infections

Two reports have identified a similar range of emotional and cognitive reactions between groups of patients with AIDS and those with malignant melanomas or acute leukaemia (Dilley et al. 1985; Temoshok et al. 1986). In addition, reports from American centres identify an ascending order of psychosocial distress from healthy homosexual men to people with AIDS to people with chronic HIV disease (Temoshok et al. 1986; Tross et al. 1986). Interestingly, it is the last group that appear to experience the highest levels of psychosocial morbidity. The conclusion from such studies is that persons with chronic HIV disease have the highest risk for psychological stress, probably due to the persistent uncertainty regarding their possible decline into frank AIDS.

Another group providing significant management difficulty is the "worried well". This group generally presents with a background history of previous psychiatric intervention, relative social isolation and conspicuous ongoing fears for personal health. In the majority, the advent of the AIDS crisis acts as a vehicle for the expression of psychological vulnerability and sexual guilt, which, in turn, leads to the expression of obsessive compulsive, depressive, and anxiety syndromes/features (Miller 1987a).

As Table 9.1 indicates, the appearance of psychological phenomena requires intervention based on recognising the patient's background and social milieu, as well as symptom diagnosis (Miller 1986). The social imperatives raising issues of self-esteem, personal control, sexuality and the potential for rejection form a crucial part of the counsellor's approach – they invariably lead to the active involvement of loved ones in longer term counselling (Miller 1987b).

Background Issues

In considering what AIDS-related counselling requires, it is important to understand what is AIDS-related counselling. In general terms, it means facilitating an understanding of the patient, or their social milieu regarding known facts about HIV, and of their responses to such facts. If one is to be a successful facilitator, it is necessary (a) to have a thorough understanding of current knowledge about HIV, (b) to have an informed understanding of patient life-styles and imperatives, and (c) to be accessible to those groups requiring and desiring intervention. In practice, this latter requirement results in AIDS counselling facilities usually being based within hospital sites offering direct medical intervention (Voldberding 1986; Gee 1986), although the strain on hospital resources together with an increased awareness of the risk of infection amongst the general population have more recently resulted in community-based (general practice and other) AIDS counselling initiatives. Similarly, cost effectiveness requires that medical and paramedical personnel are working with the necessary backup of non-medical community volunteers in providing "front line" counselling services (see Part 2).

An essential aspect of this broadening counselling initiative is the need for consistency of information. Where patients are being seen by many different community and hospital staff, most of whom perform some type of supportive and counselling function, mixed messages or variations in emphasis may easily result. If patients become confused in their understanding of "the facts", or interpretations and speculations from staff diverge, patient confidence and motivation will quickly suffer. This possibility is vitally important, as a prime aspect of counselling is to generate patient confidence in their ability to live successfully, for however long, with HIV or AIDS (Miller 1987b).

Structure of a Counselling Service

An interesting feature of the literature on AIDS-related counselling is the relative absence of examples of working counselling structures. While formats in counselling the chronically ill are the subject of many studies, referral and care plans have yet to be scrutinised empirically (Green and Miller 1987). It is important none the less to establish effective and appropriate plans of referral in order to maximise the possible benefits of counselling intervention.

One format that has been successfully established in the UK over the past four years involves two conscious assumptions: (1) the need for a team-centred approach; (2) the essential requirement for cooperation with community voluntary services. The plan of this format is illustrated in Fig. 9.1.

Although it is probable that most interventions will involve no more than perhaps two or three sessions, there is much information that must be provided and some patients will require more comprehensive follow-up, particularly where their health status interferes with their personal relationships at work, home, socially and sexually.

Clinical experience reveals the necessity for an *early* establishment of such a format, not least in order to streamline subsequent psychiatric referrals and admissions procedures for the management of suicidal and neurologically impaired patients (Miller 1987b). The growing evidence of central nervous system (CNS) disease across the range of HIV seropositives from the otherwise asymptomatic to those with frank AIDS (Cooper et al. 1985; Navia and Price 1986), together with a broadening spectrum of neuropsychiatric complications (Wolcott et al. 1985) suggests the need for all practitioners to maintain considerable alertness and a high index of suspicion at the emergence of possible signs of such dysfunction. Counsellors have a particular responsibility in this respect. Some early indications of CNS impairment are provided in Table 9.2.

Pre-test Counselling

The aim of pre-test counselling is to ensure that individuals who elect to take the HIV antibody test do so with an understanding of the basic facts about the virus, the test, and of the implications of a positive result. In many cases, the need or otherwise for antibody testing can be determined beforehand by a careful examination of the patient's sexual history and a discussion of the reasons for their fears of infection. These reasons will frequently reveal an ignorance of possible routes of transmission of HIV, such as fears of infection from shared toothbrushes, towels and linen, cooking and eating utensils and non-sexual kissing and touching.

Pre-test counselling also provides an opportunity to discuss safer sexual practices with patients, advice on risk reduction being important, regardless of the patient's

Stage 1: Presentation for testing or symptom diagnosis

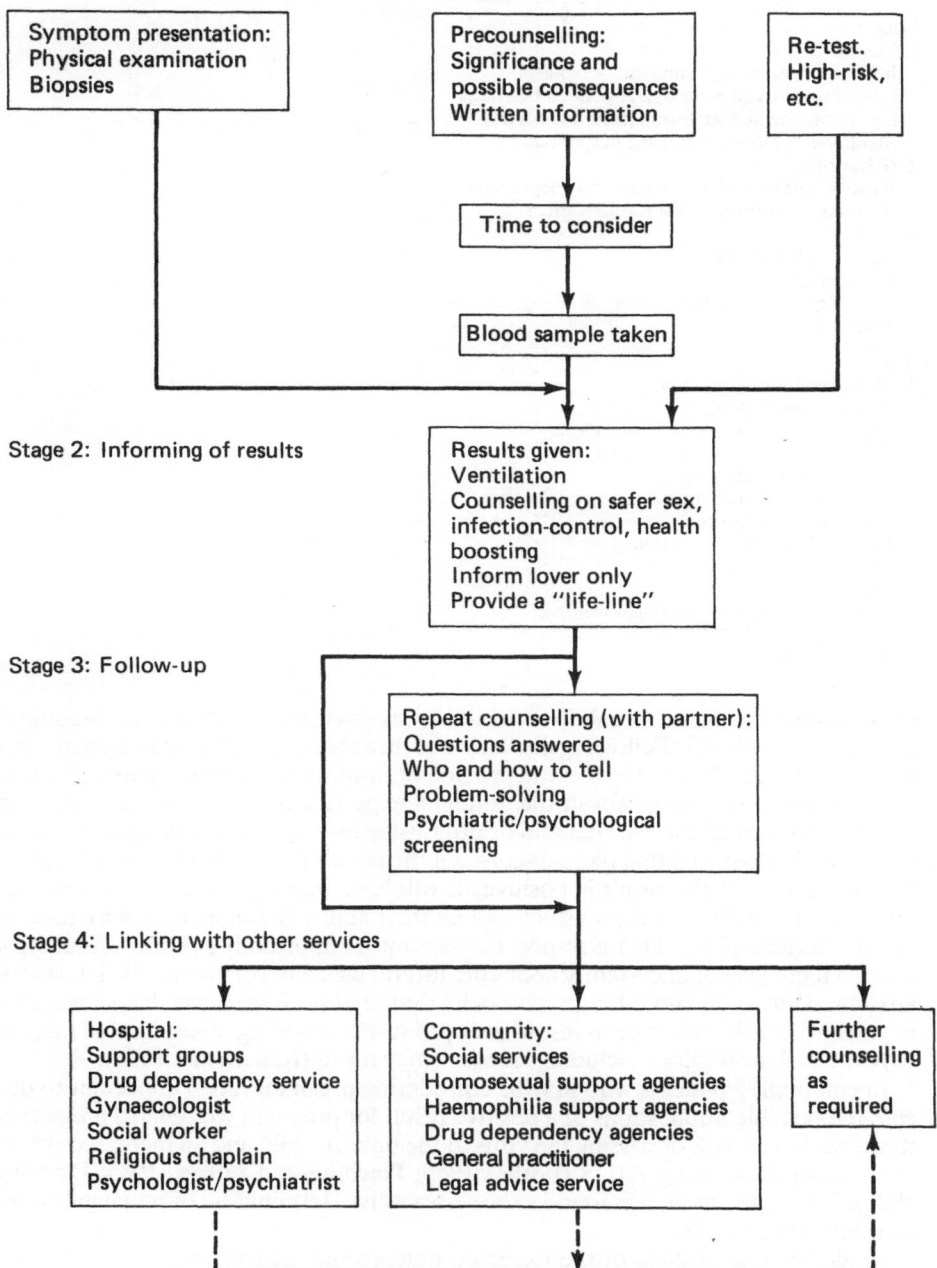

Fig. 9.1. Counselling structure for HIV-related problems.

Table 9.2. Early manifestations of HIV-related CNS involvement

Symptoms
1. Cognitive
   Impaired concentration and distractability
   Loss of memory (e.g. names, appointments, tasks)
   Disorientation and confusion (time and places)
   Mental/verbal slowing and loss of spontaneity
2. Behavioural
   Apathy, withdrawal, motoric slowing/depression
   Agitation, confusion, inappropriate affect
3. Motor
   Ataxia, unsteady gait
   Leg weakness
   Loss of coordination (e.g. eating, drinking, writing)
   Tremor and rigidity

Signs
1. Mental status examination
   Psychomotor slowing
   Impaired memory (specially short term)
   Signs of organic psychosis
2. Neurological examination
   Impaired rapid movements
   Gait ataxia (rapid turns)
   Hyperreflexia and leg weakness
   "Essential" tremor
   Dysarthria
   Impaired smooth pursuit eye movement

After Navia and Price 1986.

HIV antibody status. A breakdown of the areas discussed in pre-test counselling is provided in Table 9.3. Following discussion of these issues the patient and counsellor should establish whether the test can provide the information that the patient is seeking. It is important to establish the patient's expectation of the test outcome and whether, in view of their sexual and/or drug-using history, this is a realistic expectation. It is also essential that patients consider the possibility of a positive result before taking the test. If the result is positive, it will have many practical and emotional effects on the patient's relationships and on their ability to obtain dental treatment, medical treatment and life insurance. Perhaps most importantly a positive result produces a high level of uncertainty about the future, which may generate high levels of anxiety, depression and other psychosocial dysfunction. Experience has shown that in cases where there has been insufficient pre-test counselling, considerable rates of psychosocial morbidity – including suicide – may result (Miller et al. 1986).

In counselling patients who may be considering pregnancy, it is important to discuss the possible implications of a positive result for pregnant women. In pregnancy there is a higher risk of passing the virus to the unborn child and an increased risk of the mother developing AIDS (Forbes 1986; Pinching and Jeffries 1985; Pinching 1986). Where the mother is found to be seropositive, termination of pregnancy must be offered as an option.

In all cases, regardless of the expected outcome of the test, safer sex should be discussed in detail and in the context of the patient's own life-style (see below).

Following pre-test counselling the patient should be able to make an informed decision about whether or not to take the test and should ideally reconsider this deci-

Table 9.3. Pre-test counselling issues

The test
Is not a test for AIDS
Only indicates previous exposure to HIV
Gives no indication of prognosis, severity of infection or
    infectiousness
Sufficient time for seroconversion should have elapsed
    before the test is taken

Practical consequences of being seropositive
Ineligibility for future life insurance and some mortgages
Difficulties with obtaining dental and sometimes medical
    treatment
Exclusion from some types of employment
Dismissal, in some cases, from employment with resulting
    financial burdens

Potential psychological consequences of being seropositive
Unresolvable uncertainty
High-level anxiety, depression, guilt and obsessive
    disorders
Alteration in relationships

Other issues
All persons potentially at risk must adopt safer sex and
    risk-reduction guidelines

---

sion in the light of information given. Where pre-test counselling is being provided, the patient will often be attending in a state of extreme agitation and anxiety. This is seen most markedly in pre-counselling of the "worried well". It is therefore important that verbal information be supplemented with written information that may be more fully absorbed before a decision is finally made to test. An appointment can then be made to have blood taken a few days after the pre-test counselling session.

Safer Sex

Providing people with information about risk reduction by adopting safer sex practices is of paramount importance. Until a cure is found for AIDS or a reliable antiviral agent is found for HIV, prevention remains the only means of control. Safer sex should be discussed with all patients at risk of HIV infection, but it is important that the subject is approached with an understanding of the broader implications it may hold for an individual's social life-style. When advising patients on safer sex it is important not to present the information as a list of "dos and don'ts"; instead, emphasis should be given to the variety of sexual practices that do not constitute a risk of HIV infection. The aim of practising safer sex is to avoid any exchange of infective body fluids and thus avoid the risk of infection with the virus. In Table 9.4 the safer sex guidelines are summarised.

Having described the mode of transmission of the virus and discussed with the patient the risks attached to individual practices, it is then important to examine the patient's current sexual life-style and find ways of reducing their risk that are both realistic and acceptable. Changes that the patient feels are possible should then be discussed and the emphasis should be on reducing the risk in the long term rather than on eliminating it in the short term only. A compromise is often necessary. Patients should be encouraged to discuss safer sex with prospective partners in order to avoid situations where unsafe sex is expected. Rehearsing such situations with patients may prove helpful in building their confidence to raise the subject and to cope with possible rejection. For many homosexual men, their social life revolves around gay clubs and pubs where sex is an integral part of the social interaction. Avoiding such places can lead to social isolation; helping patients to cope with difficult situations rather than to avoid them will enable a continuation of their social life and reduce their risk of infection.

Table 9.4. Safer sex guidelines

Avoid unprotected vaginal or anal intercourse (i.e. *always* use a condom and water-based lubricant)
Avoid oral sex
Avoid "rimming" and "fisting", and *never* share sex toys (i.e. vibrators, dildos)
Discuss safer sex with all prospective partners
Safer sex is easier with fewer partners
Avoid group sex and places where sex is likely to be a pressure (e.g. "back rooms")
Mutual masturbation, frottage, massage, "dry" kissing, use of fantasy, etc. are all safe with intact skin
    contact
    There is a risk involved in all penetrative sexual activity. However, the degree of risk varies. Vaginal and anal intercourse are high-risk activities, although using a condom that does not break, tear, or come off during intercourse will provide protection. A water-based lubricant must be used when using a condom. Oral sex is a lower risk activity, particularly if no ejaculation occurs. There is no evidence that the virus is passed in saliva. Contact with saliva poses a risk of transmitting other viral infections that may be hazardous for people infected with HIV.

Post-test Counselling

Negative

Post-test counselling of patients who are negative is too easily dismissed as unnecessary. If the patient is participating in high-risk activities, it is very important to impress upon them that they have the opportunity to remain negative by following safer sex guidelines. Immediately following a negative test result, however, patients are usually very relieved – even euphoric – and their ability and motivation to concentrate may be reduced. It is useful to bring the patient back a few days after the result is given to reiterate the advice.

Positive

The aim of post-test counselling for sero*positive* patients is two-fold (a) to provide information about the virus, ways of reducing the risk of developing AIDS and infection control guidelines; (b) to support patients during the period between learning the result and coming to terms with it.

    The reactions of patients to the news that they are seropositive varies enormously. We can assume, however, that all will suffer a significant degree of shock when given the result, as the last remaining hope that they may be negative is lost. In any discussion immediately following the giving of a positive result, much of what is said will be forgotten by the patient because of the effects of shock and will need to be repeated in later sessions. It is important, however, that certain pieces of information are re-inforced at this stage. Make sure the patient understands:

1. He/she has not got AIDS but only the virus that in some people causes AIDS
2. Not everyone who is HIV antibody positive develops AIDS
3. Although they have not got AIDS, the patient is infectious and could pass on the virus.
4. There are ways in which they may be able to reduce their risk of getting AIDS, e.g. by avoiding further (sexually transmitted) infections

On learning that they are seropositive, many patients suffer from acute anxiety or depression and some may become suicidal. It is important to establish the patient's plans on leaving the clinic; where possible it should be arranged that they will not be alone and ideally there will be a close friend or loved one in whom they can confide. Most importantly they should be advised to keep the information absolutely confi-dential, with the exception of those whom they can trust absolutely. Often patients feel an overwhelming need to tell the information to outsiders (e.g. work colleagues, acquaintances, etc.). This should be very strongly advised against in the initial phase, as the repercussions arising from the "wrong people" knowing can be calamitous.

For some patients, the fear of people finding out can be difficult to cope with; inno-cent remarks can suggest that the information is already known and considerable anxiety can result. Reassuring these patients that such remarks are innocent will usu-ally be of help; in cases where people suspect, patients often have to learn to "brazen it out".

Patients should be advised to defer major decisions until they have adjusted to being seropositive. It may be helpful at this stage to let patients know that they may suffer some psychological effects over the following days or weeks. If not forewarned in this way, patients sometimes return with the added anxiety that since they learned their news they feel as if they are "going mad". It is of course important to avoid bringing on such states by suggestion. Some discussion of the physical effects of anxiety and reactive depression may also be helpful in defusing future worries that they are undergoing physical decline caused by the virus (Miller et al. 1985).

Providing a "life line" for patients is imperative. This may be a list of telephone numbers for the clinic, local AIDS information and seropositive support groups, and community telephone counsellors (e.g. the Samaritans). This is particularly impor-tant for more isolated patients who may have no-one in whom they can confide. Being seropositive can generate a higher level of anxiety than having a diagnosis of AIDS (Tross et al. 1986). Many people with a recent diagnosis of AIDS have described a sense of relief at the news – they know at last what is causing their illness and malaise.

Patients frequently report feeling a lack of control over, and uncertainty about, their future. This often results in an insatiable desire for information. Advice on safer sex and healthier living are useful means by which a sense of control can be restored. In adopting safer sex behaviour, patients are reducing their own risk of intercurrent sexually transmitted diseases, some of which may contribute to the onset of AIDS (Weber et al. 1986). At the same time, it is important for seropositive patients to realise that they can continue to have a sex life without fear of infecting their partners. Sexual contact is a source of intimacy, comfort and closeness, as well as a way of expressing affection. At a time when they may have feelings of being diseased and unattractive, sex should not be viewed as a taboo. For some patients, however, the news that they are seropositive is accompanied by a loss of libido and in such cases patients should be reassured that this is a common and temporary state.

By taking steps to maintain or improve their general health, an element of control can be restored to the patient and while it cannot provide a guarantee of avoiding AIDS it will certainly cause no harm. A healthy balanced diet with plenty of fresh foods, moderate exercise matched to the patient's physical capacity and interests and regular sleep should be encouraged. The reduction of stress, though not always easy, will also contribute to a sense of control and a healthy life-style. Patients may find areas in their lives where they can actively reduce stress; relaxation training is helpful in this context.

At this time patients will often set goals for themselves and their behaviour that are unrealistically high. It is important to encourage them gently to accept that there will be occasions when they do not live up to their expectations. Stress generated by feelings of anger and remorse that they have let themselves down or "blown" their chances of staying healthy is much best avoided!

It is important to suggest, at this time, a "sense of history" to the patient. They will not always have such an acute response to their knowledge of seropositivity, and there will come a time when they can "pigeon-hole" this news and carry on with their lives as before. Again, the confidence of the counsellor that this can be so is an important aspect in motivating patient adherence to optimal behavioural change.

Counselling Patients with AIDS, ARC and PGL

When a patient is given a diagnosis of AIDS, AIDS-related complex (ARC) or persistent generalised lymphadenopathy (PGL), the first reaction is usually one of shock associated with the implications for, and uncertainty about, the future. Following this is a need for detailed information about the diagnosis and prognosis. The many emotions associated with such a diagnosis will in some cases lead to states of severe anxiety or depression. Many practical issues arise such as dealing with implications for relationships, coping with employment problems and telling family and friends; if these are not resolved they can generate more anxiety. Open access to, and continued support from, the counsellor is imperative. When a patient is given a diagnosis of AIDS, the introduction of the counsellor should be at the earliest possible opportunity. This enables the relationship between counsellor and patient to become established and the patient thus sees the counsellor as an integral part of the management team, rather than as a sign that they may be declining psychologically.

Discussing the Diagnosis

Patients will want to discuss their diagnosis in detail. It is important that the counsellor and physician communicate well so that consistent and accurate information is given to the patient. Often patients will ask how long they have left to live, and while these questions should be dealt with honestly it is not useful to give estimates of life expectancy – they may be very inaccurate and in the meantime they can take away the patient's hopes and motivation for the future. A diagnosis of AIDS typically raises fears about death and dying and while discussion of these issues should never be forced it is important that patients appreciate that they have an opportunity to discuss these as they wish. For many patients it is not the fear of dying that is of primary concern. Rather, they may fear a slow physical decline, disfigurement, dementia and/or the effects of medication. Additionally, they may have grave fears about the ability of their loved ones to cope with the future effects of health decline. For some patients, however, the overriding fear concerns their loss of privacy and confidentiality and they should of course be reassured that this will not occur as a result of clinic or hospital policy.

Loved Ones

Lovers of AIDS patients often face graver stresses than the patients themselves. They have to cope with their own fears not only for the patient but also for the inevitable change in their relationship with the patient. Partners become carers and counsellors, while at the same time experiencing the constraints put on their relationship by the patient's physical decline. Their sexual relationship may change radically, even cease, and their social life may be considerably reduced. If the patient is unable to work there may the additional burden of increased financial pressure. If the partner too is seropositive, there will be the added anxiety for their own future health – in many instances the health decline of the lover mimics that of the patient, as though the patient were providing a model for the lover.

Partners often feel an overwhelming sense of helplessness and inadequacy. All they can do is to provide practical help and emotional support while the patient apparently continues to suffer. In some cases the strain that the illness puts on the relationship proves too much and the relationship ends. Sometimes the lover or partner may seek "permission" from the counsellor to leave the relationship. In all cases, counselling should be offered to loved ones and if they decline this it should be clear that the option remains open if they should change their minds in the future.

Families

It is often the case that when a patient is diagnosed as having AIDS their family is unaware of their homosexuality. Where families are to be informed at the patient's request, two major pieces of information are thus given at once. The counsellor has an important role in discussing with the patient when and how to approach the family with this news. It may be helpful for the patient to start by informing a sibling and eliciting their help in telling parents. Rehearsal of this process will prove invaluable beforehand.

In some cases it may fall to the counsellor to reveal the diagnosis and its implications to the patient's family directly. Before family members are brought in to counselling it is particularly important to clarify with the patient beforehand what details of their life-styles (if any) they do *not* want to share. Although many patients fear family rejection arising from previously expressed prejudices against homosexuality, experience suggests that in the majority of cases families respond well. However, as with lovers, there may be considerable strains arising from fears of the diagnosis, and families involved in the patient's care should also be offered counselling as a matter of routine.

With loved ones and family members, it is desirable where possible to secure their active involvement (subject to the patient's consent) for the following reasons:

1. To have appropriate adjustments in sexual behaviour discussed and explained
2. To have appropriate standards of hygiene explained (e.g. how to manage body fluid spillages in the home)
3. To clarify any misconceptions they or the patient may have about HIV transmission
4. To have their own and the patient's psychological responses to the diagnosis (e.g. anxiety, depression, obsessional states) explained and placed in a manageable perspective
5. To assist in effective monitoring and management of the patient's condition

Practical Issues

Many practical problems can arise following a diagnosis of AIDS. Patients may need to sort out business arrangements and make a will. Employment problems can arise from long periods off work and the subject of what to tell employers needs to be addressed. In some cases, a patient may lose his or her job and the resulting financial hardship can lead to many difficulties, even to homelessness. These problems all add to the stress of the illness and the counsellor plays an important role in referral to, and utilisation of, various statutory services designed to meet these needs.

Counsellor Support

What is often not appreciated is the *physical* impact of supporting persons in distress on the counsellor. When case loads increase with relentless urgency following media coverage of the HIV/AIDS pandemic, counselling and other staff have to face often severe increases in emotional and physical stress.

Recent studies have highlighted greatly increased levels of depression, anxiety and physical disease in physicians and other staff dealing with AIDS-related issues for more than 40% of their clinical time (McKusick et al. 1986). This morbidity is associated with the intensity of clinical involvement rather than the chronicity of involvement in AIDS.

It is important to consider these lessons during the planning stages of a counselling service, if AIDS-dedicated personnel are to avoid "burnout". The institutionalising of a staff support mechanism is an obvious safeguard. This may be in the form of one-to-one or group meetings, with confidentiality and respect for the acknowledgement of personal vulnerability being assured. Such a model for staff support has been applied successfully in other countries and other settings involving care of the chronically ill, and results in lowered staff morbidity and greater work efficiency (Gee 1986).

2. The Voluntary Sector
T. Whitehead

The two most significant risk groups for HIV infection are gay and bisexual men and people who use drugs by injection. These groups, although very diverse, share some important sociological characteristics.

The first and perhaps the most significant is the stigma attached to homosexuality and to drug use. The effect of this is to strengthen group identities and to affirm a sense of belonging at the same time as distancing the members of the group from the rest of society. Society is hostile to gay men and women and drug users. Even if individuals have not themselves experienced this hostility the knowledge of it is an important feature of the shared beliefs of the group.

The second important characteristic is the considerable ignorance in society at large about gay people and drug users. The stereotyped assumptions implicit in the use of terms such as "queer" or "junky" distort the perceptions of the individuals in the groups, leading to a lack of awareness of the diversity of life-styles, needs and expectations. Prejudiced assumptions have too often led to prejudiced treatment, both from "lay" people and from "professionals". This has understandably led in turn to suspicion from gay people and drug users towards professional services.

A third feature follows from the second. Professionals and lay people alike tend to recognise only those gay people and drug users who conform to the common stereotype; the vast majority remain hidden and unrecognised. To recognise the truth of this one has only to think of the times one hears people say that they do not know any homosexuals, even though gays are common in every community and every walk of life.

Community Organisation

Gay people's response to their perceived isolation and lack of appropriate services has been to organise a complex network of community services. These cover the spectrum from social meeting places to information and counselling services. This is not true of drug users, who have to remain hidden because of the illegality of drug use.

It is from this foundation of gay community services that the growing number of voluntary AIDS support services has evolved. This community response to the threat of AIDS is all too often ignored by the popular press more given to pointing accusing fingers at the gay community than to praising their actions. A consequence of this is that the resources offered by such groups have frequently been unrecognised by the community at large and health care workers, for whom they can be of enormous value. The common perception of AIDS as a "gay" problem and the attendant stigma, coupled with the fact that these support groups, because of their history, are gay identified, has tended further to hide their value for helping with all aspects of AIDS and HIV infection. What must be recognised is that such groups are an important resource for everyone working with, or concerned about, AIDS.

Forms of Community Support

There is now a considerable number of support groups in the United Kingdom. They tend to be one of two basic types. In practice, however, they are usually very similar.

The most significant difference is in the nature of the volunteers. Some are composed predominantly of people personally affected, usually because of infection, by HIV. These groups could therefore be considered as self-help groups, though they may well provide varying degrees of service to the wider community. Of these, the best known is Body Positive, a London-based group of HIV seropositive persons. The other type of group deliberately involves a wider selection of volunteers with the explicit aim of providing services to the general community. The Terrence Higgins Trust is the most widely recognised such group.

Other than this the differences between the groups tend to be functions of the size of the group and of local demands for services. In an area where few if any demands have yet been made, AIDS support groups, if they exist at all, may be very small with restricted services and underdeveloped skills. None the less, they all have great potential as a community resource.

The Terrence Higgins Trust

The Terrence Higgins Trust was established in 1983 and became a registered charity in January 1984. The aims of the trust are:

To provide welfare, legal and counselling help and support to people with AIDS and HIV infection, their friends and family

To disseminate accurate information about AIDS and HIV to high-risk groups, the general public and the media

To provide health education

To encourage and support research into the causes and treatment of AIDS and HIV

To facilitate and encourage the provision of care and support services for people with AIDS and HIV

The Trust is a limited company and thus has a board of directors. These directors are also sometimes called the Trustees. Their responsibility is to the proper business management of the organisation and the employment of paid staff. Day-to-day decisions on the work of the Trust and its policies are the responsibility of the Steering

Committee. The latter is made up of volunteers elected from within the Trust's working groups plus the elected offices of Chair and Vice Chair. Others may be co-opted on to the Steering Committee as required.

The Trust's Working Groups are:

Counselling – devolved into Phone Group and Buddying Group
Medical/Scientific
Health Education
Drugs Education
Social Services
Legal Services
General Purposes
Communications
Finance
People with AIDS Advisory Group
Interfaith

Some of these groups are essentially professional groups that advise the Trust on specific areas of concern. Others are composed of volunteers from a wide range of backgrounds.

The Working Groups of the Terrence Higgins Trust

The Counselling Group

The Counselling Group is one of the largest in the Trust. There are also two full-time paid staff members facilitating the work of the group.

Telephone Information Service

The telephone "hotline" service is staffed by volunteers from the Counselling Group. It operates daily from 7 p.m. to 10 p.m. and from 3 p.m. on Saturdays and Sundays.

The service receives 2000 calls per month as of August 1986. The exact number of calls varies according to press coverage of AIDS and the volume of Trust publicity and advertising. The calls cover a wide spectrum of concerns about AIDS from requests for general information through to more detailed enquiries from people afraid that they may have contracted HIV. Safer sex and drug use are frequent topics.

Telephone volunteers are trained in general non-directive counselling skills. Callers are encouraged to talk through their concerns and fears, as well as being given hard information where needed. Medical matters will be discussed in general terms, but callers with personal medical enquiries will be referred to their GP or other appropriate services.

Callers needing support or one-to-one counselling will be referred to their nearest support group. If the caller is in London, details will be taken down and referred to one of the paid counselling administrators for processing.

The Buddy Service

The Buddy Service offers one-to-one counselling and support and general home care for people with AIDS. Buddies are organised into teams covering the Greater London area. Each team is responsible to a team leader and it is his or her responsibility to allocate a Buddy when requested. As Buddying is a long-term commitment, great care is exercised in finding a suitable Buddy. Referrals come either through the telephone service and the office or direct from hospitals.

The Buddy's relationship with a person with AIDS is a very close one. It may also include helping close friends, lovers and families to cope with the illness and eventual death. Buddying is difficult and demanding work and puts great stress on the volunteers. Support for volunteers is vital and this is also organised through the Buddy teams. In addition, there are occasional courses on such issues as stress management for all Trust volunteers.

Support Groups

The Counselling Group has been administering a Support Group for people with AIDS since August 1984. It was set up so that those directly affected by the syndrome could help and support one another. The group is run by a highly qualified nurse and is held fortnightly. The Trust also encourages and supports autonomous groups for people with AIDS.

People with a positive HIV antibody result can be referred either direct to Body Positive or to the Trust's own HIV group. The Trust's HIV groups do not offer long-term support but are designed to run weekly for four weeks. They cover many aspects of learning to understand and live with HIV infection.

Medical/Scientific Group

The volunteers in this group come mainly from medical backgrounds. The group is split into a number of smaller groups which work on individual projects and meet as necessary. Most of the leading London hospitals and the medical disciplines relevant to the Trust's work are represented.

This group advises the Trust on all the medical information it publishes or otherwise requires for its work. Speakers from the group usually take part in the seminars, workshops and courses organised by the Trust. The group also maintains close links with the hospitals and medical services.

Health Education

Health education, as one of the prime functions of the Trust, encompasses many functions and responsibilities. Chief among these is the production of the Trust's leaflets on AIDS and the issues it presents, such as safer sex. Working in cooperation with other Trust groups and external agencies, the Health Education Group organises workshops and seminars for the community and for professionals concerned with AIDS prevention. The group also commissions advertising and videos and is extensively involved with the media.

Drugs Education Group

The Drugs Education Group provides support and help to people involved with drug use and to the relevant external agencies. The volunteers in this group are all professional drugs workers, experienced in working with HIV-positive drug users. The group also runs a weekly support group for HIV antibody positive drug users.

Social Services Group

This group encompasses a range of expertise from the social work services, including probation and youth services. It advises the Trust on social service entitlements of people with AIDS and the availability of these services. The other main responsibility of this group is to inform and to educate social service departments, professional organisations and schools of social work about the needs and problems of people with AIDS in the community.

Legal Services Group

The legal services group was formed to provide legal advice to people affected by HIV and to others who are at risk. They also perform research into the legal problems likely to be precipitated by HIV identification.

General Purposes Group

This group undertakes the more routine tasks of the Trust but is none the less vital to its efficient working. The group assists the Office Administrators in the day-to-day tasks of running the office, such as processing paper work and preparing mailings. The general purposes group is also responsible for servicing the computers, maintaining data banks and training volunteers in their use.

Communications Group

This group maintains communication links both within the Trust and with outside organisations and the Press. It includes amongst its volunteers a Press Officer who deals with all enquiries from the media. The group also liaises with political parties, unions and other pressure groups.

Finance Group

The finance group is responsible for raising money for the Trust so that it can continue to expand to meet its objectives. A large proportion of the funding comes from individual donations and charity benefits. The Trust also makes extensive use of deeds of covenant, which are supervised by the group. A large proportion of funding for administration comes from the Department of Health and Social Services and other statutory bodies. The Treasurer is a member of the Finance Group and is responsible for helping each group set and keep to its budget.

People with AIDS Advisory Group

People with AIDS are not regarded as simply a "client" group. Individuals have long played an important role as volunteers. Through this group, they are enabled to have considerable impact on the running of the Trust and its policy decisions.

Interfaith Group

AIDS has thrown up many religious and spiritual dilemmas, both for people with the syndrome and those working in the field. It was in response to these needs that the Interfaith Group was established. Volunteers in this group come from a range of Christian and Jewish backgrounds. Many of them are ordained and work full time in the ministry.

The group provides information on religious and spiritual matters for the Trust's work. Members of the group also provide counselling for clients requiring spiritual guidance. An important part of the group's work is to try to improve the response of the established churches to the problems of AIDS in the community. To achieve this, the Trust has organised and taken part in seminars and conferences for the clergy and laity.

Training

All volunteers to the Trust go through a comprehensive training programme. The initial part of the programme consists of a general training day, covering the medical and social aspects of HIV infection and the basic elements of the Trust's counselling approach. More intensive training then follows according to the needs of the groups in which volunteers elect to work, the most rigorous being for those groups providing direct service to the public, especially the counselling group.

Telephone information volunteers follow a sequence of five training evenings. These evenings make use of role play, discussion, outside speakers and written material. At the start of these sessions, each trainee is allocated a supervisor who will support them and monitor them through training. These are held at weekly intervals and the programme is as follows.

Week 1. An introduction to counselling skills

Week 2. Medical review of HIV as it affects injecting drug users, haemophiliacs and heterosexuals

Week 3. Structures and referral systems operating in phone work

Week 4. Calls from people with HIV infection

Week 5. Safer sex

Once the trainees have completed all five training sessions they are then assessed. If they are suitable they can then begin phone training.

The initial programme is five sessions on the phone service accompanied by a fully trained and experienced volunteer. On each shift some aspects of the practical components of being on the phone are covered, as well as discussion of the calls. If satisfactory progress has been made the volunteer will then be asked to take on phone duties.

Preparing for working as a Buddy follows the same pattern. However, Buddy training is much more practical, with emphasis on preparing for death and bereavement counselling.

Body Positive

Body Positive is a self-help group for people who are HIV antibody positive. Its structure is less formalised than that of the Terrence Higgins Trust and more easily open to participation. HIV antibody-positive people coming to the group for help and advice are encouraged to become active members not clients. This open approach is one of the great strengths of Body Positive.

Services

Body Positive provides a wide range of counselling and support services to people with HIV. These range from formal one-to-one counselling from trained volunteers through to very informal but welcome opportunities for socialising with other antibody positive men and women. The counselling service can be arranged by prior appointment, as in the Terrence Higgins Trust, or people may drop in to one of the regular open meetings that the group hosts. The opportunity to walk into such a meeting is a most valuable service. Many people come to Body Positive through direct referral from sexually transmitted disease (STD) clinics on receipt of a positive antibody result.

Body Positive also operates a telephone line where callers are able to speak to other HIV antibody-positive people. Until the group has its own office this phone service is available only through referral from other agencies such as the Trust.

Structure

Body Positive, in keeping with its informal approach, has only two permanent officers. These are the Secretary and the Treasurer. The two main working bodies of Body Positive are the Policy and Resources Group and the Counselling Group. These groups both seek the active involvement of HIV antibody-positive people. In this way there is a very real sense in which people can say that Body Positive is "our group".

Policy and Resources Group

The main responsibilities of this group are to formulate policy and to coordinate services. Planning and fundraising also come under the remit of this group.

Counselling Group

The counselling group is responsible for the provision of counselling services, including the telephone service. The group also coordinates training for volunteers. Other Body Positive activities such as the regular social evenings and health promotion campaigns are coordinated through ad hoc groups represented on the Policy and Resource group.

Future plans for Body Positive include a permanently staffed "drop in" and resources centre in West London.

Summary for Local Planning

The Terrence Higgins Trust and Body Positive are described as not so much models for local action but examples of the services that need to be provided. It is hoped that local communities will choose those aspects most applicable to local needs and resources and adapt as necessary.

Planning of services is best undertaken as a joint initiative between relevant groups already in existence, such as gay community counselling services, and the District Medical Services. It should also involve the Health Education Service and Social Services.

Both the Terrence Higgins Trust and Body Positive are happy to give all the help and advice they can in establishing local support and education services.

References

Christ GH, Wiener LS (1985) Psychosocial issues in AIDS. In: DeVita VT, Hellman S, Rosenberg SA (eds) AIDS: Etiology, diagnosis, treatment and prevention. JB Lippincott, Philadelphia, pp 275–298

Cooper DA, Gold J, Maclean P et al. (1985) Acute AIDS–retrovirus infection; definition of a clinical illness associated with seroconversion. Lancet ii: 537–540

Dilley JW, Ochitill HN, Perl M, Volberding PA (1985) Findings in psychiatric consultations with patients with acquired immune deficiency syndrome. Am J Psychiat 142: 82–85

Forbes PB (1986) The significance of AIDS in obstetric practice. Br J Hosp Med 33: 342–346

Gee G (1986) Nursing. In: Jones P (ed) Proceedings of the AIDS conference 1986. Intercept, Newcastle upon Tyne, pp 131–146

Green J, Miller D (1987) The psychosocial impact of AIDS. In: Gottlieb MS, Jeffries JD, Mildvan D et al. (eds) AIDS 1. John Wiley, Chichester, pp 287–302

McKusick L, Horstman D, Abrams D, Coates T (1986) The impact of AIDS on primary practice physicians. Paper presented at the IInd international conference on AIDS, Paris, 23–25 June 1986.

Miller D (1986) Psychology, AIDS, ARC and PGL. In: Miller D, Weber J, Green J (eds) The management of AIDS patients. Macmillan Press, Basingstoke, pp 131–149

Miller D (1987a) Psychosocial issues in HIV disease and the worried well. In: Miller D, Weber J, Green J (eds) The management of AIDS patients, 2nd edn. Macmillan Press, Basingstoke, in press

Miller D (1987b) Living with AIDS and HIV. Macmillan Press, Basingstoke

Miller D, Green J (1986) Counselling for HIV infection and AIDS. In: Pinching AJ (ed) Clinics in immunology and allergy: AIDS and HIV infection, vol 6, part 3. WB Saunders, London, pp 661–683

Miller D, Green J, Farmer R, Carroll G (1985) A "pseudo-AIDS" syndrome following from a fear of AIDS. Br J Psychiat 146: 550–551

Miller D, Jeffries DJ, Green J, Harris JRW, Pinching AJ (1986) HTLV-III: should testing ever be routine? Br Med J 292: 941–943

Navia BA, Price RW (1986) Central and peripheral nervous system complications. In: Pinching AJ (ed) Clinics in immunology and allergy: AIDS and HIV infection, vol 6, part 3. WB Saunders, London, pp 543–558

Pinching AJ (1986) The spectrum of human immunodeficiency virus (HIV) infection: routes of infection, natural history, prevention and treatment. In: Pinching AJ (ed) Clinics in immunology and allergy: AIDS and HIV infection, vol. 6, part 3. WB Saunders, London, pp 467–488

Pinching AJ, Jeffries, DJ (1985) AIDS and HTLV-III infection: Consequences for obstetrics and perinatal medicine. J Obstet Gynaecol 92: 1211–1217

Temoshok L, Mandel JS, Moulton JM et al. (1986) A longitudinal study of AIDS and ARC in San Francisco: preliminary results. Paper presented at the annual meeting of the American Psychiatric Association, Washington DC, 13 May 1986

Tross S, Holland J, Hirsch D et al. (1986) Psychological and social impact of AIDS spectrum disorders. Paper presented at the second international conference on AIDS, Paris, 23–25 June 1986

Volberding PA (1986) Clinical care. In: Jones P (ed) Proceedings of the AIDS conference 1986. Intercept, Newcastle upon Tyne, pp 119–130

Weber JN, Wadsworth J, Rogers LA et al. (1986) Three-year prospective study of HTLV-III/LAV infection in homosexual men. Lancet i: 1179–1182

Wolcott DL, Fawzy FI, Pasnau RO (1985) Acquired immune deficiency syndrome (AIDS) and consultation-liaison psychiatry. Gen Hosp Psychiat 7: 280–292

Chapter 10

AIDS and Homosexuality in Britain: A Historical Perspective

J. Austoker

A significant feature of the current AIDS epidemic is the association that has been made between sex and disease. The fact that the transmission of the virus in developed countries has occurred in the main in male homosexuals, as well as in intravenous drug users, both groups that are already the subject of deeply rooted hostilities and discrimination, has had a profound impact on the social, political, economic, and even clinical responses to the disease.

Underlying much of the fear invoked by AIDS is the strong belief that there is a causal link between the disease and certain sexual practices and social habits associated with homosexuals (Weeks 1986). From this perspective there has been a tendency to shift gradually to the idea that homosexuals cause the disease and, finally, to the notion that homosexuality itself is a disease. It is the reputation of AIDS as a disease of homosexuals, a "gay plague", that has most sharply focused society's responses. It is because of this association with a group that is deemed to have violated society's expectations of what is acceptable sexual behaviour that AIDS has acquired a particular stigma, a symbolic significance (Sontag 1983; Weeks 1986). The social and cultural reactions to AIDS, the public bewilderment, the fear and the panic, can be seen as a response not just to the terrifying clinical realities of the disease itself, but to what we believe sex ought to be and, accordingly, how sexuality should be regulated and controlled (Weeks 1985). The identification of AIDS with homosexuals has thus uncovered deeper concerns about homosexuality, serving to undermine and to challenge the whole basis of a gay identity. This tendency to focus on individual sexual conduct and to blame the victims for actually causing the disease is consistent with the traditional attitudes associated with venereal diseases in Britain in the nineteenth and early twentieth centuries. This is clearly reflected in the language, the imagery, and the rhetoric of the current crisis, and very often in the proposed solutions.

Initially, and alarmingly, AIDS was seen as nature's "awful retribution" upon the gay community (Buchanan 1983). Some went so far as to suggest that gays should be rounded up and quarantined. This spectre of coercion and the implied violation of civil liberties bring to mind a striking historical parallel with the association made in the nineteenth century between female prostitution and the incidence of the venereal diseases, and the subsequent passing in England in the 1860s of the Contagious Diseases (CD) Acts, which enforced compulsory inspections, in certain garrison towns, of women suspected of being carriers (Ware 1969; McHugh 1980; Walkowitz 1980). A combination of feminists and social purity crusaders, with the Christian feminist Josephine Butler prominent, campaigned vigorously against the "double standard" of morality enshrined in the series of CD Acts passed between 1864 and 1869 (Bristow 1977; McHugh 1980). Aimed at protecting the health of the armed forces, this exceptional legislation, which was overseen by the Admiralty and the War Office, specified that, in particular garrison towns and ports, a special body of plain clothes policemen should have sweeping powers to detain those suspected of being prostitutes, submit them to a statutory medical examination and, if necessary, compel them to be interned in a lock hospital for a period of up to three months, where they would be treated and also put to work. The enforcement of the Acts provided enormous scope for blackmail, corruption and above all victimisation.

The remarkable feature of these Acts was the bitter opposition they engendered. This is important in the present context with respect to AIDS, where mandatory screening of high-risk groups, premarital testing, universal surveillance of the entire population, and the quarantine of all HIV antibody-positive individuals, are subjects of heated debate, however illogical or inappropriate such solutions might appear to be. The opponents of the CD Acts condemned them for demoralising the state by recognising vice, for the discrimination against women implicit in the "double standard", for infringing individual rights, or for violating a "divinely ordained order" by which venereal disease punished illicit intercourse. Between 1870 and 1885 at least 520 books and pamphlets on the subject were written by those seeking the repeal of the Acts, 17367 petitions bearing 2606429 signatures were presented to the House of Commons, and more than 900 public meetings were held (Sigsworth and Wyke 1980). The concerted attacks led to the suspension of the Acts in 1883, and their repeal in 1986. The emphasis of the campaign changed from a concern with rescuing the fallen prostitute to ensuring a single standard of chastity and moral purity for both sexes.

Numerous purity organisations were formed in the 1870s and 1880s. An attack on sexual incontinence followed, based on the belief that sexual desire was a moral and intellectual error that should be controlled "by the will and good faith" (Bristow 1977). To the moral crusaders, the syndromes of schoolboy masturbation, public school immorality (homosexuality) and prostitution were all inextricably linked with one another (Wilson 1885; Lyttelton 1887).

A series of scandals, such as that in Dublin Castle in 1884, and the Cleveland Street scandal of 1889–90, focused attention on sexual decadence, and strengthened the link between homosexuality and prostitution (Chester et al. 1976). In general, however, it is likely that changing attitudes to homosexuals were unintended consequences of other major changes. An important factor here was the emphasis in the social purity campaigns on the dangers of male lust. What is noteworthy is that the social purity campaigners of the 1880s saw both prostitution and male homosexuality as "the products of undifferentiated male desire", and it is significant that the three major enactments between 1885 and 1912 that affected homosexuals were all aspects of a wider moral restructuring, and were in fact primarily concerned with female prostitution (Bristow 1977; Weeks 1977). This was to become a familiar phenomenon. Even in the 1950s, a single committee – the Wolfenden Committee – was set up to consider both homosexuality and prostitution.

The changing legal situation and the emergence of new conceptualisations of homosexuality were accompanied by, and intimately associated with, its "medicalisation". The tendency of the medical profession to regard homosexuality as a pathological condition provided both theoretical explanations of "criminal behaviour" and the provision of potential solutions for such "sickness" or "mental illness". Thus, any notion of the immorality of homosexual activity became linked with scientific theories and explanations that formed boundaries against, and within which, homosexuals were forced to define themselves. In this way the "medical model" profoundly influenced the individualisation of homosexuality.

From a social perspective, it was not so much the prosecutions of homosexuals but the moral panic that accompanied certain celebrated cases that was to become significant in creating a public image of homosexuality. The famous "Three Trials" of Oscar Wilde in 1895 served to focus attention on what was deemed to be acceptable or deviant sexual behaviour. Against such deviance was the fact that the family had become the symbol of stability and purity. The institution of the family was, claimed Josephine Butler (1885), in accordance with the law of God. It was lust that threatened the family and the nation.

The purity leagues launched a massive campaign to disseminate their message to the young and their efforts constitute the origins of sex education. They produced vast numbers of anti-sexual books, pamphlets, leaflets and lantern shows, succeeding in terrifying young people for over 30 years on such issues as the evils of masturbation. The belief was that the young, taught self-control, would not resort to prostitutes later. It was part of a wider concern to order their lives and socialise them into being good citizens. By the early years of this century the horrors attributed to sexual vice included the declining birth rate and the contribution of the alleged increase in the venereal diseases to racial degeneration (Scharlieb 1916).

The campaign to combat the venereal diseases was rooted more in anxiety about moral degeneration, racial inefficiency, and social decline, than in a concern about the diseases themselves (Tredgold 1918). It was promiscuity that became the major issue rather than the control of venereal diseases. The medical profession too began to draw upon the changing sexual and social attitudes, joining purity advocates in the effort to educate for chastity. This paved the way for the social hygiene movement of the early twentieth century. Purity was seen as essential for national preparedness. Social purity workers joined with socialists, imperialists, feminists, Christian leaders, politicians, eugenists and medical practitioners in an alliance patriotically concerned with the racial destiny of the British people.[1]

Some eugenists responded to the threats of biological deterioration by calling for campaigns against alcohol and venereal diseases (VD). Others, however, argued against preventive medicine, maintaining that keeping alive those "types" susceptible to crime, insanity, alcoholism and the contagious diseases would only be jeopardising the innocent. Arnold White, an ardent exponent of negative eugenics, felt that the state had no duty to treat these "avoidable" diseases, declaring that "the Empire will not be maintained by a nation of out-patients" (White 1901).

The International Conference on Prostitution and Venereal Diseases, held in Brussels in 1899, highlighted the contribution made by VD to the problems of public health. In the same year the British Medical Association unsuccessfully urged the government to institute an enquiry into the prevalence of VD in the civilian population (May 1946). Representations on the subject were made to the Prime Minister by the Interdepartmental Committee on Physical Deterioration in 1904, and the Poor Law Committee in 1909, who referred in its Report to the terrible havoc caused by VD to the health of the community. The Eugenics Education Society (EES) became increasingly concerned with the issue and in 1910 began urging the Local Government Board (LGB) for the provision of increased treatment facilities.[2] At the same time, the National Council for Public Morals, under the guidance of James Marchant, was conducting ambitious programmes for "the regeneration of the race, spiritual, moral and physical". In its widely publicised manifesto of 1911, it expressed alarm at the incidence of VD, insisting that the young "should be taught to entertain high conceptions of marriage, as involving duties to the future of the nation and the race".[3]

The women's movement was equally interested in promoting moral instruction to prevent VD and in breaking down the "conspiracy of silence" on the subject (Bland 1982). From Louisa Martindale's *Under the Surface* (Southern Publishing Co., Brighton, 1908) and Christabel Pankhurst's famous tract on VD, *The Great Scourge and How to End It* (E. Pankhurst, 1913), in which she referred to VD as the cause of "physical, mental and moral degeneracy and of race suicide . . . ravaging the community", to Louise Creighton's more moderate *The Social Disease and How to Fight It* (Longman, Green & Co, London 1914), there was a call for widespread education for sexual purity and moral hygiene.[4]

Under increasing pressure, the LGB commissioned a report on VD in which it was argued that there was an urgent need to extend the provision of facilities for diagnosis and treatment (Johnstone 1913). The subject was finally brought to public attention when, on 22 July 1913, the *Morning Post* published a letter, signed by 38 prominent doctors, demanding the appointment of a Royal Commission on Venereal Diseases. In the following month at the International Medical Congress in London, the subject of VD took a leading place at discussions with Paul Ehrlich, the discoverer of Salvarsan, as "the hero of the congress".[5] The congress passed a resolution to press the Government for a Royal Commission.

The culmination of all this medical and public pressure was that the government finally acquiesced and set up a Royal Commission on 1 November 1913. Many of the leading doctors active in securing the appointment of this commission were included in its membership. The remaining members determined to form an organisation to direct public opinion on the importance of the problem and, in January 1914, Major Leonard Darwin outlined to the EES Council a scheme for the formation of a national committee to be concerned with the prevention of VD.[6] In November of that year, the National Council for Combating Venereal Diseases was established under the Chairmanship of Sir Thomas Barlow, President of the Royal College of Physicians.[7]

Despite the presence on its various committees of many of the most eminent members of the medical profession, the council decided to wage its propaganda campaign for VD control on social and moral rather than on medical grounds.[8] Entrenched in the nineteenth-century elitist assumption that social progress depends on moral factors, it desired to reform the morals of the nation. Beginning with the conviction that the foundation for social and racial hygiene was biological, it sought to stimulate biological and physiological instruction, with sex hygiene as an integral part (Austoker 1981).

The First World War was to have a profound impact on sexual activity, and thereby on the incidence of VD, which, according to some commentators, reached "terrifying proportions" (Prinzing 1916; Morris 1917; Harrison 1922). Masturbation, homosexuality and prostitution once again came under close scrutiny (Hirschfeld 1934). In the prison camps, masturbation was to become "a favourite pastime, later degenerating into a mania, with all its evil, physical and moral consequences" (Fischer and Dubois 1937). In the trenches, alliances "contracted between men" played a role "of some little importance in the heroic conduct of the combatants". Indeed, at its most extreme, the war was regarded fearfully by some as "one huge brothel in which sexual maniacs could indulge their mania to their heart's content, and were very often decorated and promoted for their bestiality" (Fischer and Dubois 1937).

Such attitudes aside, it was female prostitutes who became the focus of attention in the campaign to combat VD (Neville-Rolfe, 1925, 1935, 1949). Apportioning blame to women for the spread of VD was by now a familiar phenomenon. One of the first tasks of the NCCVD was to consider the advisability of making the conscious transmission of VD a criminal offence. In November 1914, shortly after the inaugural meeting of the council, Sir Thomas Barlow suggested to the Home Office that it should consider enacting legislation that would empower the police to exclude prostitutes from military areas (Buckley 1977). The Home Office referred the matter to the War Office, but both were reluctant to sponsor a measure of this sort for fear of stirring up a controversy similar to that over the CD Acts. Sir Ernley Blackwell, Assistant Under-Secretary in the Home Office, was worried that such legislation would pave the way for the "amateur" prostitutes, "a great army of girls . . . over whom the police have no control".[9] It was this "army" he believed, which was the true source of the spread of the diseases. The Home Office went so far as to claim that 75% of all VD came from the "enthusiastic amateur". As far as the NCCVD was concerned, the presence of these amateurs formed "a serious menace, for these girls are often ignorant of the dangers of their mode of life" (May 1916). To the Home Office, the way to tackle the "amateur" was to encourage the women patrols and make use of the women police (Bland 1986). Many feminists concerned with the issues of moral degeneracy had joined volunteer police units during the war. The women patrols, founded in October 1914 by the National Union of Women Workers and headed by Louise Creighton, an active member of the Executive Committee of the NCCVD, saw their role as "moral watchdogs", in existence to "restrain the behaviour of women and girls who congregate in the neighbourhood of the camps" (Allen 1925). By 1918, the voluntary women's patrols had become a section of the Metropolitan Police, and by 1923 had been granted full powers of arrest. Included in their duties was the need to investigate sex offences. This task was also carried out by local vigilance committees, who kept watch for the obscene and the indecent. Thus the Public Morality Council employed a special officer in London in order to observe public behaviour, in particular prostitution and male homosexual importuning.

The government was coming under increasing pressure from various sources to introduce some form of regulation of prostitution. By the middle of 1915, the Canadian military were agitating strongly for reform measures, maintaining that both military and moral efficiency were at stake (Buckley 1977). They obtained support from various sources, including the English Home Commanders, the Archbishop of Canterbury and the NCCVD. As far as the War Office was concerned, action against women was a matter for the Home Office. The National Council concentrated its efforts accordingly. It suggested to the Home Secretary, Herbert Samuel, that an Act of Parliament should be passed making it an offence for a person knowingly and deliberately to do anything calculated to transmit the disease. The council presented a draft bill to Samuel, suggesting that, if the future of the race was at stake, even sexual intercourse between man and wife should be liable to conviction if the "offender" knew that he or she had VD.[10]

Under increasing pressure from both the Colonial Office and the War Office, the Criminal Law Amendment Bill was introduced into the House of Commons by Sir George Cave, the new Home Secretary, on 15 February 1917. It stipulated that it was a crime for anyone suffering from VD in a communicable form to have sexual intercourse with another person.[11] Its repressive clauses were widely condemned by women's organisations, but, equally, it was ridiculed by those who desired a return to the CD Acts. The latter saw the bill as "half-hearted" and virtually unenforceable (Buckley 1977). With diseased women now reputed to be "swarming" around the troops, the Canadians threatened to take steps if the troops were not adequately protected from these diseased women "lying in wait for clean young men who come to give their lives for their country" before "the future of our race [is] damaged beyond any comprehension or conception" (Buckley 1977). Finally, in 1918, the government introduced the highly controversial and much reviled regulation 40D under the Defence of the Realm Act (DORA), making it a summary offence for *any* woman with VD to "have sexual intercourse with any member of HM Forces, or solicit or invite any members to have sexual intercourse with her".[12] Despite the strong emphasis on the need to "control" or to "regulate" the prostitute, it was impossible to implement legislation aimed at an ill-defined social group. Moreover, 40D aroused hostile public sentiment, with angry opposition from feminists, the lay and medical press, the Trades Union Congress and many members of both Houses of Parliament.

The NCCVD remained firmly convinced that the basis of the problem was far more fundamental, being deeply rooted in complex social and moral issues. Just as certain measures have been rejected by some in the current AIDS crisis because they are viewed as potentially encouraging homosexuality, so the NCCVD refused to countenance any form of medical prophylaxis, such as disinfection, feeling that sanctioning its use would only legitimise "irregular" and promiscuous sexual behaviour (Towers 1980).[13] With 54 884 soldiers hospitalised by VD in 1917 the Government had approved a policy to issue condoms to the forces. The NCCVD was horrified, convinced that this would hasten the end of the morals and attitudes of Victorian England. The only way to cope with the problem in the long term, it felt, was to encourage "moral prophylaxis".

The Royal Commission on Venereal Diseases had reported in 1916, after examining a substantial and authoritative array of opinions on sexuality and disease (White 1920). By this time, the NCCVD had been frightening the public for two years with horrendous images of VD. As the incidence of VD continued to rise, social purity tried to halt the progress. The Royal Commission Report revealed how widespread the incidence of the disease had become. Moreover, the report revealed strong class differences in access to treatment – while Salvarsan was readily available to the rich, the poor were often refused admission to hospitals and could lose their entitlement to insurance benefits (Weeks 1981). The commission steadfastly refused to tackle the question of medical prophylaxis, despite the prophylactic advantages of condoms or postcoital disinfection, fearing that this would hasten the decline in sexual mores. Any move to make sexual activity "safe" would, it believed, provide "a blow to the country's morals" (Scharlieb 1924).

The commission concluded that:

The importance of wisely conceived educational measures can hardly be exaggerated . . . especially should instruction and warning be given to the young of the moral and physical dangers which may imperil them.[14]

The government accepted and implemented the main recommendations of the Royal Commission. Under the terms of the Public Health (VD) Regulations of 1916, the

LGB issued, to local authorities, directives for the provision of treatment. This led to the establishment of state-financed pathological laboratories for diagnosis, free supplies of Salvarsan to medical practitioners, and the development of free facilities for treatment. Local authorities were encouraged to undertake the establishment of clinics and to enlighten the public in the causes and treatment of VD. This, therefore, avoided the controversial issue of prophylaxis and set the scene for an extensive campaign to develop sex education.

In accordance with the recommendations of the Royal Commission, the government recognised the NCCVD as *the* body to coordinate the activities of all the organisations dealing with the subject of VD.[15] This meant that, as a semi-official organisation, it received financial aid from the LGB and, from 1919 to 1942, received grants from central and local government and submitted to the Ministry of Health estimates of expenditure to pursue its extensive propaganda campaign.[16]

On the publication of the Royal Commission's findings, its Chairman, Lord Sydenham of Combe, a former army officer and high imperial official, became the first president of the NCCVD and all but one of its commissioners became members, several being elected to its executive committee. Lord Sydenham called for education measures – a "great national crusade" – promoting the purity of life by religious and moral teaching, thereby encouraging "temperance, chivalry and healthy exercise."[17]

After the Armistice, DORA, including 40D, was revoked, but the interest of the NCCVD in the "amateur", and the whole question of promiscuity, continued with an expansion in its educational work. W. H. Fisher, President of the LGB, urged the council in 1918, "do all that you can to spread this wonderful educational propaganda which you are carrying on. Carry the torch of medical and moral truths right out amongst the young people. . . ."[18]

The education campaign mounted by the council reached an enormous audience and attempted to shape public understanding of venereal diseases. By 1920, the council had delivered a phenomenal 8560 lectures to a civilian population of nearly 2500000.[19] Lectures were supplemented by the sale or distribution of pamphlets and books published by the council. These covered such subjects as *Local authorities and the problem of VD*, *What mothers must tell their children*, *England's girls and England's future*, *The duty of knowledge*, and *The incidence of VD in relation to school life and school teaching*. Leaflets and posters were widely distributed, coupled with an extensive advertising campaign to raise funds. As the magnitude of the task increased, so did the council's sense of urgency. The rising prevalence of VD during the war acted as an incentive to the council to expand the scale of its campaign. As has been the case more recently with AIDS, syphilis was perceived as the link between the morally respectable and the socially deviant. The council sponsored showings of the controversial play, *Damaged Goods*, which related the slow destruction of a French family by congenital syphilis.[20] The play showed to packed houses. In one week in Darlington it was seen by 5000 people and in Newcastle and Wakefield, hundreds of people were turned away. At the end of the war, the council resorted to what Edward Bristow has called "repulsive scare tactics", by exploiting a new propaganda vehicle, "the VD horror film" (Bristow 1977). Under pressure from the council, local authorities waived the prohibition on sexual realism in films to allow VD films to be shown throughout the country to youth groups, mothers' meetings, factory workers and large public gatherings. The film *End of the Road*, with sequences about the danger of contracting VD from a soldier's kiss, praised the good work of the women's voluntary patrols. It contained scenes that were "horrible and revolting" and many people were reported coming out of the cinema "in a state of coma", yet the crowds flocked to see it.[21] At South Shields, four cinema showings accompanied by lectures were attended by 6600 people, but nearly 10000 were turned away.[22] In Darlington, it was seen by an audience of 4000, in Gateshead by 30000, in Hull by 9000, and in London it ran at the Poly Cinema in Regent Street for six weeks to full houses. In York, a lantern lecture on *Love, Marriage and Parenthood*, followed by films on *Maternity* and *The Girl Who Doesn't Know* were attended by 4300, and 25000 saw *Damaged Goods* in Hull.

While the propaganda campaign mounted by the NCCVD was extensive, and was seen or heard by considerable numbers of people, its impact is much harder to assess. What is known is that, at the end of the war, the incidence of VD was still rising. In general, it can be said that the nature of the propaganda expounded by the council greatly enhanced prejudice and ensured the continuation of the stigma attached to the venereal diseases. By blaming certain ill-defined groups in society for their weakness, lack of moral fibre, and promiscuity, and by considering them as contaminated and corrupt, the council in fact discouraged rather than encouraged attendance at approved treatment centres. Its education campaign was based on the somewhat simplistic assumption that the problem could be solved if individuals would only act more responsibly in their sexual relationships (Brandt 1987). The campaign lacked compassion, was largely sensationalist, and had very little impact on public health. With respect to AIDS, where there is an obvious role for health education in helping to reduce irrational fears, and perhaps in slowing down the spread of the epidemic, the NCCVD's education campaign provides salutary lessons by demonstrating approaches that should at all costs be avoided. At the same time the council's policy on prophylaxis remained one of complete opposition, based on the conviction that prophylaxis served to legitimise promiscuous sex (Towers 1980). A Ministry of Health departmental committee, reporting in 1919, endorsed this belief by stating that the issuing of packets for self-disinfection was neither desirable nor practicable, stressing that priority should be given to moral teaching and early treatment. Moreover, the council's policy was reinforced by legal restrictions in that the 1917 VD legislation made it an offence to advertise remedies for diseases arising from sexual intercourse.

Alarmed at the sort of propaganda being expounded by the NCCVD, the Society for the Prevention of Venereal Diseases (SPVD) was set up in October 1919. It aimed to disseminate knowledge of the prophylactic means of combating VD by the use of officially approved disinfectants. It strongly advocated the use of the so-called "packet system". Over a number of years it tried unsuccessfully to influence the Ministry of Health to approve this alternative strategy by removing the legal obstacles to the retailing of these disinfectants, which varied from Metchnikoff's Formula, based on calomel and carbolic acid, to Condy's Fluid, a solution of potassium permanganate (Wansy Bayly 1935). But despite the vigorous attack mounted against it, the NCCVD continued to receive the staunch support of the Ministry of Health. Neville Chamberlain, Minister of Health in 1923, and a former president of the Birmingham Branch of the NCCVD, was particularly unsympathetic to the views of the SPVD, rejecting all suggestions advocating prophylaxis, despite the recommendations to the contrary of the Trevethin Committee, which reported in 1923 (Ministry of Health 1923).

The widespread occurrence of VD posed major challenges to the NCCVD. Rather than look towards the development of a well-organised and comprehensive VD service, it chose to tackle the problem largely through the medium of sensational and inappropriate education. In doing so, it was promiscuity rather than VD that became its main target.

The failure to control in particular gonorrhoea gave no grounds for complacency. This escalating problem was in part a reflection of the existing treatment regimens. Treatment after the First World War was "in its crude and sometimes dangerous infancy" (Adler 1980). By the mid 1920s the recommendations of the Royal Commission were being implemented by the NCCVD, but there was little reason for optimism or satisfaction. Despite the rapid spread of VD clinics, these were, on the whole, characterised by a lack of finance and hampered by the failure to introduce an organised and coordinated service. Moreover, the clinics were often sordid and dingy, serving to amplify the stigma attached to the diseases (Harrison 1941; Adler 1982).

The Second World War gave an impetus towards the development of a more coherent service, but the notorious Regulation 33B introduced a major setback (Shannon 1943). Under this legislation, a compulsory medical examination could be performed on any person mentioned by at least two others as a possible source of infection. This once again invoked memories of the CD Acts and demonstrated the impracticality of introducing compulsory legislation of this nature.

The substantial rise in the incidence of VD infections from the mid 1950s onwards provided the basis for a renewed onslaught on promiscuity. A BMA report in 1964 on *Venereal Disease and Young People* suggested a great increase in promiscuity that paralleled the rise in VD. Indeed trends in the incidence of VD, showing considerable increases in the number of cases of the prevailing infections, were considered to be sensitive indicators of the level of promiscuity[23] (Nicol 1963). It did not go unnoticed that the proportion of infections amongst homosexuals was "remarkably high"[24] (Jefferies 1956; Schofield 1973).

Public prejudice against homosexuals and considerable anxiety about homosexuality had pervaded public debate in the early 1950s, culminating in 1954 in the much-publicised and sensational trial for homosexual offences of Lord Montague of Beaulieu and Peter Wildeblood, Diplomatic Correspondent of the *Daily Mail* (Wildeblood 1957). The trial exposed the need for an enquiry into the question of homosexual offences. Accordingly, the Interdepartmental Committee to investigate prostitution and homosexuality offences was set up in 1954 under Sir John Wolfenden.

In the face of rapidly altering psychiatric and medical attitudes to homosexuality, the Wolfenden Committee found itself confronted by a series of paradoxes and apparent contradictions (Home Office, Scottish Home Department 1957). Thus, while it considered that homosexuality could not legitimately be regarded as a disease, it believed that in certain cases "associated psychiatric abnormalities" occurred. The acceptance of an underlying psychological basis for both homosexuality and prostitution meant that such concepts as "treatment" and "cure" were not totally dismissed. The committee acknowledged and "greatly regretted" the loosening of moral standards, and deplored the potential damage promiscuity might bring to the "basic unit of society", the family. Yet at the same time it articulated principles that were to provide a pragmatic basis for approaching questions of moral regulation. This opened the way for the limited but none the less significant permissive legislation of the late 1960s, and even provided the framework for the more substantial proposals on morality that came in the 1970s (Weeks 1977, 1981).

By the 1960s "permissiveness" had become a "political metaphor" (Weeks 1981). A series of reforms in the late 1960s of the laws relating to various aspects of sexual behaviour transformed opportunities for the young and therefore must be regarded as the most significant legislative changes on morality since 1912. This was inevitably accompanied by new waves of fears and anxieties. The sexual liberation movement that emerged in the late 1960s, initially in the United States and by the early 1970s in Western Europe, was the product of a variety of forces. The unifying feature was the emphasis on taking control over one's own life and one's own body. There was a growing public awareness of homosexuality and a limited and uneven acceptance of the gay identity. "Gay" was taken to suggest "a new defiance of moral norms", a coming out, the emergence of a "gay pride" (Weeks 1981).

This euphoria was to be short lived. Prejudice was deep seated and homosexuals never fully escaped social ostracism. The arrival of AIDS instantly exploded the myth of gay liberation. With almost indecent haste there was a transition from "gay pride" to "gay plague". Once again the irresistible combination of disease and sexuality provided the basis for victim-blaming as homosexuality and AIDS became firmly intertwined. AIDS was viewed not first as a disease but as a symptom of a far more profound sociosexual deviance. It was a punishment for the sexually maladjusted and the promiscuous. Very quickly homosexuals were regarded as "the guilty, the major force in spreading the disease to innocent bystanders" (Brandt 1987). Many of the characteristics of the earlier anti-venereal diseases campaign resurfaced, but with one important and frightening difference. The high mortality and the lack of any cure or any vaccine set AIDS apart from all other sexually transmitted diseases. In its unremitting onslaught on promiscuity and sexual mores, the NCCVD had been privileged in that it could make choices between various strategies and options. With AIDS there are fewer choices and potentially a far higher penalty to pay.

The assumption that AIDS is the consequence of sexual misbehaviour pervades much discussion and assessment of the problem. Despite the clear lessons of the past, many continue to believe that a disease such as AIDS should be regulated not on scientific or medical grounds but rather by recourse to moral control. The idea that the sexually transmitted diseases are products of individual behaviour and should therefore be controlled through modifications of individual conduct only serves to reduce the emphasis on external determinants of disease and health (Brandt 1987). With AIDS, while it is recognised that changes in sexual behaviour may well have a significant impact on the spread of the disease, the experience of the earlier anti-venereal diseases campaigns should have served to highlight the inappropriateness of draconian measures such as quarantine, and instilled into society the needs for tolerance and compassion. The stigma that continues to be attached to AIDS victims must be seen as an extension of the moralising and the values that have been associated with venereal diseases throughout this century.

The history of venereal diseases raises a central problem in the conflict it creates between what is good for the individual and what is right for the health of the community. There is a sharp tension between individual freedom, or responsibility, and society's needs. Calls for dramatic, legally enforceable, coercive measures such as segregation and confinement of those identified as having AIDS or carrying it bring to mind the dreaded CD Acts of the 1860s, where such direct legal measures were tried but abandoned to a chorus of hostile protest (Porter 1986). If we are to learn anything from that experience it is only that compulsion is unlikely in the long run to be a constructive or satisfactory policy. In the absence of a vaccination or an effective therapeutic regime, a programme of health education is a positive way forward, although by no means a solution. The sensational and misdirected campaign run by the NCCVD does not provide a suitable model. What it taught us was that propaganda promoting a particular moral position and aimed primarily at promiscuity and not at the disease itself only enhanced prejudice and intolerance, added to existing stigma, and did nothing to reduce the incidence of the disease or in any way to improve public health. To overcome the irrational and hysterical fears surrounding AIDS a coordinated effort is demanded, whereby the government, medical professionals, health educators, the victims and the general public combine forces to conquer this threat to society.

AIDS poses major challenges to clinical and basic biomedical sciences as well as to medical and health professionals. Much remains to be established about the natural history of HIV infection and the interaction between this agent and its host. Ultimately it will be fundamental research that will provide a basic understanding of the immunological, epidemiological, clinical and psychological factors needed to effect greater control of the disease. Until such time, it will be necessary to overcome innate prejudice, rise above apocalyptic statements and frightening statistics, and implement a national strategy incorporating long-term planning to control the disease. The role of constructive education programmes aimed at altering sexual practice but at the same time avoiding moralising and victimisation has only recently been realised. Although most health educators readily concede that it will take far more than education to change sexual habits, such a programme needs to be implemented in order to inform how the virus is spread, and how best to protect against getting it. Such advice, should, where possible, be based on common sense, use explicit language and targeted messages, and be placed in the hands of those with the necessary expertise. This would constitute a level of sex education never dreamed of by those pioneer advocates of sex instruction, the NCCVD. But then the stakes are higher and the challenges far more formidable.

Notes

1 Shield (October 1913).

2 EES, Fourth annual report (1911–1912) pp 22–23.

3 Prevention (National Council of Public Morals, July 1911).

4 Shield (April 1914).

5 Eugenics and venereal disease, Eugenics Review (1913–1914) 5: pp 264–270. This refers both to Medical Congress and the LGB Report.

6 EES Council minutes (16 January 1914).

7 The pamphlet, *National Council for Combating Venereal Diseases* (NCCVD, 1914), includes details of the aims and objectives of the Council, membership and speeches delivered at the inaugural meeting on 11 November 1914. Also see NCCVD, Provisional Executive Committee Minutes (23 November 1914).

8 See NCCVD, First annual report (1914–1916), pp 2–7 and 37–39.

9 See the evidence of Sir Ernley Blackwell to the *Joint Select Committee of the House of Lords and the House of Commons on the Criminal Law Amendment Bill and the Sexual Offences Bill* (1918), p 187.

10 NCCVD, Parliamentary Committee Minutes (16 November 1916). The Appendix includes the draft of "An Act to Prevent the Dissemination of Venereal Diseases". Sir Thomas Barlow and Sir Malcolm Morris presented this to Herbert Samuel on 17 November 1916. See NCCVD, Second annual report (1916–1917),

p 26. Also see NCCVD, Legislative Committee minutes (2 November and 9 November 1916). A second deputation visited the new Home Secretary on 17 January 1917.

11 See Parliamentary debates, House of Commons, XCIII (1917), pp 61–179.

12 For the attitude of the NCCVD to 40D, see NCCVD Third annual report (1917–1918), pp 30–33.

13 The *British Medical Journal* gave considerable coverage to the debate. In 1920 the National Council of Public Morals undertook an enquiry into the medical and moral aspects of the argument, including a detailed enquiry into the advantages of disinfection. See Report on the prevention of venereal disease (William and Norgate 1921).

14 Report of the Royal Commission on Venereal Diseases (HMSO, 1916), Cd 8189, paras 213–229.

15 Cd 8189 (see n. 14 above), paras 225–226, 236.

16 British Social Hygiene Council, Combating venereal disease. A statement of the services placed at the disposal of Local Authorities (16 July 1930).

17 Lord Sydenham of Combe, The Work of the National Council. In: NCCVD, First annual report (1914–1916), pp 12–17.

18 W. Hayes Fisher, The place of legislation in the campaign against venereal disease. In: NCCVD, Third annual report (1917–1918), p 15. For the full discussion, see ibid. pp 11–23.

19 NCCVD, Fifth annual report (1919–1920), p 8. Also see NCCVD Propaganda Committee minutes.

20 NCCVD, Medical Committee minutes (30 October 1916).

21 Shield (December 1919 and January 1920).

22 NCCVD, Fourth annual report (1918–1919), p 115.

23 Venereal diseases in England and Wales, Extract from the Annual Report of the Chief Medical Officer for the year 1961, Br J Vener Dis (1963) 89: 41–46; for the year 1963, Br J Vener Dis (1965) 41: 41–45.

24 Extract from the Annual Report of the Chief Medical Officer for the year 1961, Br J Vener Dis 39: 41.

References

Adler MW (1980) The terrible peril: a historical perspective on the venereal diseases. Br Med J 281: 206–211

Adler MW (1982) History of the development of a service for the venereal diseases. JR Soc Med 75: 124–128

Allen M (1925) The pioneer policewoman. Chatto & Windus, London

Austoker J (1981) Biological education and social reform. The British Social Hygiene Council 1925–1942. MA dissertation, University of London

Bland L (1982) "Guardians of the race" or "Vampires upon the nation's health?": female sexuality and its regulation in early twentieth century Britain. In: Whitelegge, E, Arnot M, Bartels E et al. The changing experience of women. Martin Robertson, Milton Keynes, pp 373–388

Bland L (1986) In the name of protection: the policing of women in the First World War. In: Brophy J, Smart C (eds) Women-in-law. Routledge, Kegan & Paul.

Brandt AM (1987) No magic bullet. A social history of venereal disease in the United States since 1880. Oxford University Press, New York

Bristow EJ (1977) Vice and vigilance. Purity movements in Britain since 1700. Gill and Macmillan, Dublin

Buchanan PJ (1983) New York Post, 24 May

Buckley S (1977) The failure to resolve the problem of venereal disease among the troops in Britain during World War I. In: Bond B, Roy I (eds) War and Society, vol 2. Croom Helm, London, pp 65–85

Butler J (1885) The Sentinel, April: 441

Chester L, Leitch D, Simpson C (1976) The Cleveland Street affair. Weidenfeld & Nicolson, London

Fischer HC, Dubois EX (1937) Sexual life during the World War. Francis Aldor, London

Harrison LW (1922) A sketch of army medical experience of venereal disease during the European War, 1914–1918, 2nd edn. NCCVD, London

Harrison LW (1941) The present trend of incidence of venereal diseases in England and Wales, and methods of control. Br J Vener Dis 17: 249–256

Hirschfeld M (1934) The sexual history of the World War. Panurge Press, New York

Home Office, Scottish Home Department (1957) Report of the Committee on Homosexual Offences and Prostitution. HMSO, London

Jefferies FJG (1956) Venereal disease and the homosexual. Br J Vener Dis 32: 17–20

Johnstone RW (1913) Report on venereal diseases (Cd 7029) HMSO, London

Lyttelton E (1887) Causes and prevention of immorality in schools, 2nd edn. GT Purnell, Croydon

May O (1916) The prevention of venereal diseases in the army. NCCVD, London

May O (1946) British Social Hygiene Council: its origin and development. British Social Hygiene Council, London

McHugh P (1980) Prostitution and Victorian social reform. Croom Helm, London

Ministry of Health (1923) Report of the Committee of Inquiry on Venereal Disease. HMSO, London

Morris M (1917) The nation's health. The stamping out of the venereal diseases. Cassel & Co Ltd, London, New York, Toronto and Melbourne

Neville-Rolfe S (1925) The relationship between venereal disease and the regulation of prostitution. British Social Hygiene Council, London

Neville-Rolfe S (1935) Sexual delinquency. In: Llewellyn-Smith H (ed) The new survey of London life and labour, vol IX. PS King and Sons, London

Neville-Rolfe S (1949) Prostitution. In: Neville-Rolfe S (ed) Social biology and welfare. George Allen & Unwin Ltd, London, pp 163–219

Nicol CS (1963) Venereal diseases. Moral standards and public opinion. Br J Vener Dis 39: 168–172

Porter R (1986) History says no to the policeman's response to AIDS. Br Med J 293: 1589–1590

Prinzing F (1916) Epidemics resulting from wars. Clarendon Press, Oxford

Scharlieb M (1916) The hidden scourge. C Arthur Pearson Ltd, London

Scharlieb M (1924) Reminiscences. London

Schofield M (1973) The sexual behaviour of young adults. Allen Lane, London

Shannon NP (1943) The compulsory treatment of venereal diseases under regulation 33B Parts I and II. Br J Vener Dis 19: 22–33; 67–77

Sigsworth EM, Wyke TJ (1980) A study of Victorian prostitution and venereal disease. In: Vicinus M (ed) Suffer and be still. Women in the Victorian age. (Paperback edition) Methuen & Co. Ltd, London, pp 77–99

Sontag S (1983) Illness as metaphor. Penguin, Harmondsworth

Towers BA (1980) Health education policy 1916–1926: venereal disease and the prophylaxis dilemma. Med Hist 24: 70–87

Tredgold AF (1918) Mental deficiency in relation to venereal disease. NCCVD, London.

Walkowitz J (1980) Prostitution and Victorian society. Cambridge University Press, Cambridge, London, New York, Melbourne, Sydney

Wansy Bayly H (1935) Triple challenge or war, whirligigs and windmills. Hutchinson & Co., London

Ware HRE (1969) The recruitment, regulation and role of prostitution in Britain from the middle of the nineteenth century to the present day. Ph.D. thesis, University of London

Weeks J (1977) Coming out. Homosexuality politics in Britain from the nineteenth century to the present. Quarter Books, London, Melbourne, New York

Weeks J (1981) Sex, politics and society. The regulation of sexuality since 1800. Longman, London and New York

Weeks J (1985) Sexuality and its discontents. Routledge & Kegan Paul, London

Weeks J (1986) Sexuality. Ellis Horwood and Tavistock, Chichester and London

White A (1901) Efficiency and empire, Methuen, London

White D (1920) Synopsis of the Final Report of the Royal Commission on Venereal Diseases. NCCVD, London

Wildeblood P (1957) Against the Law. Penguin, Harmondsworth

Wilson JM (1885) Sins of the flesh. Social Purity Alliance, London

Subject Index

Acalculous cholecystitis 147
Acquired immune deficiency syndrome, *see* AIDS
Acyclovir 49–50, 89, 106, 153
Adenine arabinoside 85
AIDS 1, 2, 3, 5, 8, 10, 12, 15, 45, 52, 54, 59, 65, 92, 129–98
  adult cases reported 131
  and gay liberation 194
  as disease of homosexuals 185
  challenges to clinical and basic biomedical sciences 195
  changing prevalence 133–4
  clinical aspects 138–57
  cost of care 138
  counselling and support 163–83
    background issues 165–6
    community organisation 175–6
    community support 176
    diagnosed patients 172
    discussing the diagnosis 172–3
    families 173–4
    hospital and statutory services 163–75
    planning of local services 182
    post-test counselling 170–2: negative patients 170; positive patients 170–2
    practical issues 174
    pre-test counselling 166–9
    pregnancy counselling 168
    psychological/psychiatric phenomena 164–5
    safer sex issues and guidelines 168–70
    structure of 166
    Terrence Higgins Trust 176–82
    voluntary sector 175–82
    see also Body Positive, *and* Terrence Higgins Trust
  counsellor support 174–5
  cryptosporidosis in 70
  definition 129
  diarrhoea 71–2, 143
  dysphagia 143
  early indications of CNS impairment 166
  epidemiology 129–38
  first reports of disease 129
  gastrointestinal complications 143
  health education in 192
  hepatic complications 143
  heterosexual transmission 133
  high-risk groups 132
  historical perspective 185–98
  immunological parameters 141
  Isospora belli in 70
  malabsorption 143
  management 151–7
  Microsporidia in 71
  minor opportunistic infection 140
  natural history of infection 135
  neuropsychological complications 147–8
  prevalence and control 136–8
  prognostic markers for disease progression 140
  pulmonary complications 142–3
  retrosternal discomfort 143
  stage I 139
  stages II, III and IVa 139–42
  stages IVb–e 142–57
  threat of 8–9
  weight loss 143
  years of potential life lost (YPLL) 131
  see also HIV infection
AIDS-associated virus (ARV) 129
AIDS-related complex (ARC) 172
Alcohol 187
Alpha interferons, *see* Interferons
Amoebiasis, *see Entamoeba histolytica*
Amoebic dysentry 61
Amoebic species 65
Amoxycillin 126
Ampicillin 126

Anogenital warts, *see* Genital warts
Antivirals   156–7
ARA-AMP   85, 88
ARC, *see* AIDS-related complex
ARV, *see* AIDS-associated virus
Autoimmune chronic active hepatitis   79
Autonomic nervous system dysfunction   46
Azathioprine   91
Azidothymidine (AZT)   155

Bacterial infections   15–39, 155–6
Bath houses   8, 9, 10
Benzathine penicillin   125
Blastocystis hominis   72
Body Positive   12, 181–2
Bowenoid papulosis   104
Bowen's disease   104
Brachyspira aalborgi   29–30
Branhamella   143
Bushke-Löwenstein tumour   104

California   9
Campylobacter cinaedi   33
Campylobacter fennelliae   33
Campylobacter fetus   32
Campylobacter fetus ssp. *fetus*   32
Campylobacter fetus ssp. *jejuni*   32–3
Campylobacter infections   32–4, 147
Campylobacter intestinalis   32
Campylobacter jejuni   32
Campylobacter trachomatis   34
Candida   140, 143, 147, 156
Carcinoma of the penis   104, 105
Cautery   107
CD4 antigen   138
Cell-mediated immunity (CMI)   101
Cellular immunity   138
Central nervous system (CNS)   147, 166
Cerebral toxoplasmosis   155
Cervical cancer   41
Cervical cytology   101
Chlamydia psittaci   24–7
Chlamydia trachomatis   24–7
Chlamydial infections   24–7
   clinical aspects   24–6
      pharyngeal infection   26
      rectal infection   25
      urethral infection   24–5
   diagnosis   26–7
   treatment   27
   TWAR   24
Chloramphenicol   31
Chloroquine   67
Cholestasis   147
Cleveland Street scandal   186
Clindamycin   70, 155
CMI, *see* Cell-mediated immunity
CMV   145, 146, 148, 150
CNS, *see* Central nervous system
Codeine phosphate   155

Colistin   31
Computer-assisted tomography (CT) scan
   147–8
Condyloma acuminatum   101, 105
Condylomata acuminata   102, 103, 107
Condylomata lata   104
Condy's Fluid   192
Contagious Diseases (CD) Acts   186, 194
Corneal infection   71
Corticosteroids   155
Co-trimoxazole   154, 155
Cryotherapy   107
Crypt abscesses   26
Cryptococcal meningitis   148
Cryptococcus neoformans   147
Cryptosporidium   59, 68–70, 143, 145
   clinical features   69
   diagnosis   70
   in AIDS patients   69, 70
   life-cycle   68
   organism   68
   pathogenesis   69
   prevalence   68–9
   transmission   69
   treatment   70
Cryptosporidiosis   155
CT, *see* Computer-assisted tomography
Cytomegalovirus   52–4, 78, 142, 145, 146, 148,
   150
   clinical features   53
   diagnosis   53
   epidemiology   53
   in AIDS   54
   treatment   54

Defence of the Realm Act (DORA)   190, 191
Deoxyacyclovir   89
"Desacralisation" of sex   11
Dexamethasone   155
Diathermy   107
2',3'-Dideoxynucleoside analogues   156
Dientamoeba fragilis   65
Disseminated disease   71
DNA   41–2, 49, 52, 81, 84, 85, 89, 100, 104, 105
DNA polymerase   81, 85, 88, 89, 92
Doxycycline   27, 125
Dublin Castle scandal   186
Dyskeratosis   101

Emotional partnership   10–11
Encephalitozoon cuniculi   71
Endolimax nana   65
Entamoeba coli   65
Entamoeba hartmanni   65, 72
Entamoeba histolytica   59–65
   clinical significance   61–3
   cysts and cyst carriage   61–4
      treatment   64
   in AIDS patients   65

incidence 59–60
invasive amoebiasis 61–2
    diagnosis of 62
    treatment 64
prevalence in homosexual men 60
serological tests 62
transmission among homosexual men 60–1
treatment 64
zymodeme classification 63–4
Epstein–Barr virus (EBV) 78, 103, 140
Erythromycin 27, 125
Erythromycin stearate 34
Erythroplasia of Queyrat 104
Escherichia coli 34
Eugenics Education Society 188

Fidelity without sexual exclusivity 10–11
First World War 188
Fluconazole 156
Fordyce's syndrome 103
Fungal infections 156
Furazolidone 71
Fusidic acid 31

Ganiciclovir (DHPG) 154, 155
Gay community 10
    relationships in 10–12
Gay liberation movement 7
Gay Men's Health Crisis 12
Genital herpes 41–52
    advice and counselling in 51
    biosynthesis 42
    clinical features, first episode 45–6
    complications 46
    diagnosis
        clinical 47
        laboratory 47–9
    differences between primary and recurrent 47
    epidemiology 42–4
    incidence 43–4
    latency 44–5
    natural history 44–5
    prevention 52
    psychological events following first attack 51
    reactivation 44–5
    recurrent 46–7
    treatment 49–50
        evaluation in controlled trials 50
        recurrent infections 50
    virology 41–2
Genital warts 99–109
    anus and anal canal 103
    associated infections 102
    clinical features 102–4
    complications 104–5
    concomitant infections 106
    diagnosis 105–6
    differential diagnosis 103–4
    epidemiology 102

extent of disease 106
histological features 100
history 99–100
immunology 101
incidence 102
infectivity 102
medical treatment 106
oral cavity 103
pathology 100–1
penis and scrotum 102–3
premalignant and malignant disease 104
surgical treatment 107–8
treatment 105–8
virology 100
Giant condyloma 104
Giardia lamblia 59, 65–8
    clinical significance 67
    diagnosis 67
    pathogenesis 67
    prevalence in homosexual men 65–6
    treatment 67
Gonorrhoea, *see Neisseria gonorrhoeae*
Group B streptococci 143

Haemophilus influenzae 143
Hepatitis 77–98, 147
    categories of 77–8
    chronic active 84
    chronic persistent 84
    diagnosis 78–9
    management 79–80
    serological tests 79
Hepatitis A virus 77
Hepatitis B immunoglobulin (HBIG) 93–4
Hepatitis B virus infection 77, 80–95
    active immunisation 94–5
    acute 81
    antiviral therapy 85–92
    chronic 81, 85–92
    interactions with HIV 92–3
    natural history of 81–4
    prevention 93–4
    structure of virus 80–1
Hepatitis C virus 78
Hepatitis delta virus (HDV) 78
Herpes simplex virus (HSV) 41–5, 52, 145, 148,
    153
    cultures 47–8
    HSV1 42, 43, 45, 48–9
    HSV2 42, 43, 45, 48–9
Herpes zoster infection 148, 153
Hirsutes papillaris penis 103
HIV 70, 92, 105, 106, 130, 133, 148, 150, 186
    classification system 130
    HIV1 138
    HIV2 139
    interactions with hepatitis B virus 92–3
HIV infection 15, 52, 65, 87, 101, 103, 129
    and syphilis 127
    chronic 139–42, 145, 152

HIV infection – *cont.*
  complications of   152
  cost of care   138
  dissemination in   154
  management   151–7
  natural history of   135
  primary defect induced by   138
  therapy of   153
  see also AIDS
Homosexuality
  as disease   5
  as psychosocial condition   4
  as way of life   8
  attitudes to   1, 2
  development of   6
  fundamental facts of   4
  historical perspective   185–98
  history of   2
  intentional promotion of   7
  medical model of   7
  natural or perverse   4
  origins of   3
  professional approach to   5
  remedicalisation of   9
  see also Lesbianism, *and* Male homosexuality
Homosexuals
  characterization of   5
  identity of   5–6
  modern   12
HPA23   156
Human immunodeficiency virus, *see* HIV
Human leukocyte interferon   85, 88
Human papillomavirus (HPV) infection   100–8
  see also Genital warts
Human T lymphotropic virus type III
      (HTLVIII)   129
9-(2-Hydroxy-1-(hydroxymethyl)
      ethoxymethylguanine)   54

Idoxuridine   106
Inguinal lymphadentis   34
Interferons   88, 106, 107
Interleukin-2   155
International Conference on Prostitution and
      Venereal Diseases   187
Intestinal spirochaetes   30
Intestinal spirochaetosis   29
Iodamoeba buetschlii   65
Isospora belli   59, 70–1, 143
  in AIDS patients   70
  incidence   70
  life-cycle   70
  treatment   71

Jarisch–Herxheimer reaction   126

Kaposi's sarcoma (KS)   129, 146, 150–1, 153
Kernig's sign   46

Ketoconazole   156
Koilocytes   101

Laser therapy   108
Legislation against sodomy   3
Lesbianism   1, 4–6, 8, 10, 12
Local authorities   7
Local government   7
Local Government Act (1988)   7
Local Government Board (LGB)   188, 191
Loperamide   155
Lymphadenopathy-associated virus (LAV)   129
Lymphoblastoid interferon   88
Lymphogranuloma venereum (LGV) infection
      25–7
Lymphoma   151

Male homosexuality
  cultural perspectives   1–13
  liberalisation of laws affecting   3
  regulation of   3
  see also Homosexuality
Meningitis   46, 122
Meningoencephalitis   71
Meningovascular neurosyphilis   122
Mepacrine   67
Metchnikoff's Formula   192
Metronidazole   31, 67
Microsporidia   59, 71, 143
Minocycline   27
Molluscum contagiosum   54, 104
Moral obligation   11
Movements, emergence of   6–8
Mycobacterium tuberculosis   143, 146–7, 155
Mycoplasma   27–8
Mycoplasma genitalium   27
Mycoplasma hominus   27–8

National Council for Combating Sexual Diseases
      (NCCVD)   189–95
Neisseria gonorrhoeae   15–22, 34, 99, 102, 106,
      153
  aetiology   15–18
  clinical features   18
    pharyngeal infection   19
    rectal infection   18
    urethral infection   18
  diagnosis   19
  distribution of serogroups   17
  sites infected with   16
  strains infecting homosexual men   17
  treatment   19–22
    pharyngeal infection   20–2
    rectal infection   20
    results of   22
    urethral infection   19–20
Neisseria meningitidis   23
  anorectal colonisation   24

Neisseria meningitidis – cont.
  pharyngeal colonisation 23
  urethral colonisation 23–4
Neurosyphilis 122–3
New York 6, 7, 9
Next of kin 12
Non-condylomatous wart virus infection of the
    uterine cervix 101
Nosema 71

Oxytetracycline 27

Papovaviridae 100
PCP, *see Pneumocystis carinii* pneumonia
Penicillin 30, 31, 125, 126
Pentamidine 154, 155
Persistent generalised lymphadenopathy
    (PGL) 135, 139, 172
Phosphonoformate 54, 154
Pityrosporum 156
Plato's Retreat 9
Pneumocystis 154
Pneumocystis carinii pneumonia (PCP) 129,
    141, 142, 150, 155
Podophyllin resin 106
Podophyllum emodi 106
Podophyllum peltatum 106
Prednisolone 91
Promiscuity 8, 9
Prostitution 9, 186, 189
Protozoal infections 59–75, 154
Public Health (VD) Regulations (1916) 190
Pyrimethamine 71, 155
Pyrimethamine-sulphadoxine 154

Quinine 70, 155

Rectal spirochaetosis 28–31
  inflammatory changes in rectal mucosa 31
  prevalence of 29
  source of 30
  spirochaetes associated with 29
  treatment 31
Relationships
  attitudes to 9–12
  in gay community 10–12
Reverse affirmation 5
Reverse transcriptase inhibitors 156
RNA 42, 77, 78
Rosaramicin 27
Royal Commission on Venereal Diseases 188,
    190, 193

Safer sex practices 12
Salmonella 147, 155
Salmonella enteritidis 34
Salmonella paratyphi 32

Salmonella typhi 32
Salmonellosis 32
San Francisco 6–8
Sandstone Sex Commune 9
Science of sex 4
Seborrhoeic dermatitis 156
Secularisation of sex 4
Self-help groupings 12
Sexual activity 10–11
  frequency of 9
Sexual anarchy 10
Sexual behaviour 194
Sexual identity 6
Sexual minorities 7
Sexual needs 9–10
Sexual practices 10
Sexual promiscuity 8, 9
Sexual relationships 11
Sexual revolution 11
Sexual tradition 7
Sexuality
  atttudes to 9–12
  history of 2
  relations between true and false 2
Shigella boydi 31
Shigella dysenteriae 31
Shigella flexneri 31
Shigella sonnei 31
Shigellosis 31–2
  diagnosis 32
  histology 32
  intestinal symptoms 32
  treatment 32
Society for the Prevention of Venereal Diseases
    (SPDV) 192
Sodomy, legislation against 3
Spectinomycin 126
Spiramycin 70, 155
Spirochaetes 28
Steroid therapy 89
Streptococcal infections 34
Streptococcus pneumonia 143
Sulphadiazine 71, 155
Sulphamethoxazole 71
Suramin 156
Surrogate families 11
Syphilis 99, 102, 111–28
  and HIV 127
  causative organism 113–14
  classification of 115
  clinical aspects 114–24
  epidemiology 111–13
  incidence of 111–13
  latent 121
    diagnosis 121–2
    natural history 123–4
  neurosyphilis 122–3
  primary 115–19
    diagnosis 117–19
    differential diagnosis 118–19
  secondary 119–20

Syphilis – *cont.*
    diagnosis   120
    differential diagnosis   120
  serological tests   118, 124
  treatment   125–6

Terrence Higgins Trust   12, 176–82
  Buddy Service   178
  communications group   180
  counselling group   177
  drugs education group   179
  finance group   180
  general purposes group   179
  health education functions   178
  interfaith group   180
  legal services group   179
  medical/scientific group   178
  people with AIDS advisory group   180
  social services group   179
  support groups   178
  telephone information service   177
  training programme   181
  working groups of   177–82
Tetracyclines   27, 31, 125, 126
Tinidazole   67
Toxoplasma gondii   147

Transverse myelitis   46
Treponema pallidum   104, 113–14, 125
Trevethin Committee   192
Trimethoprim   71
Trimetrexate   155

Ureaplasma   27–8
Ureaplasma urealyticum   27–8
Urinary tract infections   34

Vanomycin   31
Venereal Disease and Young People   193
Venereal diseases   187–95
Venereal warts, *see* Genital warts
Viral infections   41–58

Warts, *see* Genital warts
West Hollywood   8
Whipple's disease   146
Wilson's disease   79
Wolfenden Committee   187, 193

Zidovudine   92, 155–7